Marine Life
A Quick Read

Seals
Leopard Seal
Harbor Seal
Harp Seal
Hooded Seal
Fur Seal
Ring Seal
Walrus
Sea Lion

Sharks
Hammerhead Shark
Great White Shark
Nurse Shark
Bull Shark
Tiger Shark
Thresher Shark
Blue Shark
Mako Shark
Megalodon

Jellyfish
Lion's Mane Jellyfish
Moon Jellyfish
Cannonball Jellyfish
Barrel Jellyfish
Box Jellyfish
Portuguese Man of War

Reef Fish
Clown Fish
Yellow Tang
Angelfish
Pufferfish
Lionfish
Parrotfish

Crustaceans
Lobster
Hermit Crab
Decorator Crab
King Crab
Shrimp
Krill
Whales
Orca
Blue Whale
Beluga Whale
Narwhal
Humpback Whale
Sperm Whale
Minke Whale
Dolphins
Bottlenose Dolphin
Dusky Dolphin
Spinner Dolphin
Australian Snubfin Dolphin
White Sided Dolphin
Rays
Cownose Ray
Giant Reef Manta Ray
Spotted Eagle Ray
Round Stingray
Giant Stingray
Turtles
Leatherback Turtle
Green Turtle
Flatback Turtle
Hawksbill Turtle
Loggerhead Turtle
Game Fish
Marlin
Swordfish
Sailfish
Blue Fin Tuna
Tarpon

Marine Life

Copyright © 2024 JFBonham
All rights reserved. No part of this book may be reproduced, distributed, or transmitted in any form or by any means, including photocopying, recording, or other electronic or mechanical methods, without the prior written permission of JFBonham, except in the case of brief quotations embodied in critical reviews and certain other noncommercial uses permitted by copyright law.
Unauthorized use or reproduction of this book may result in legal action.

Seals

Leopard Seal

The leopard seal is a magnificent marine creature that inhabits the icy waters surrounding Antarctica. Its appearance is striking, with a sleek, elongated body and a coat adorned with distinctive spots and patterns, much like its namesake, the leopard. These seals are well adapted to their cold environment, where they can often be found near pack ice and icebergs, blending in seamlessly with their surroundings.

In terms of diet, the leopard seal is a top predator, preying on a variety of marine life such as fish, squid, krill, and even other seals. Their powerful jaws and sharp teeth enable them to catch and consume their prey with ease. Despite their solitary nature, leopard seals are known for their territorial behavior, especially during the breeding season when conflicts between individuals may arise.

One of the most interesting aspects of leopard seals is their impressive swimming ability. They are agile and swift swimmers, capable of reaching speeds of up to 25 miles per hour in the water. This makes them highly effective hunters, employing tactics like "porpoising," where they leap out of the water to surprise their prey from above.

Despite their formidable reputation, leopard seals also exhibit curiosity towards humans. They have been observed approaching divers and boats, displaying a mix of caution and inquisitiveness. This behavior highlights their intelligence and adaptability in interacting with their environment.

Female leopard seals are notably larger than males, with some individuals reaching lengths of up to 12 feet and weighing over 1,000 pounds. This size difference is a characteristic feature of many seal species and plays a role in their reproductive strategies and social dynamics.

Leopard seals play a vital role in the Antarctic ecosystem, contributing to the balance of marine populations and helping regulate prey species' numbers. Their presence is indicative of a healthy marine environment and underscores the interconnectedness of all life forms in this remote and harsh region.

In summary, the leopard seal is a captivating marine predator with a distinctive appearance, well-suited to its icy habitat. Its diet, behavior, and unique adaptations make it a key player in the Antarctic ecosystem, showcasing the wonders of nature in one of the most extreme environments on Earth.

Harbor Seal

The harbor seal, also known as the common seal, is a charming marine mammal found along coastlines in the Northern Hemisphere, from the icy waters of the Arctic to the temperate regions of Europe, Asia, and North America. These seals are known for their sleek, torpedo-shaped bodies, which are perfect for navigating through water with ease. Their coloring varies from light gray to brown, often with darker spots or markings scattered across their fur. This camouflage helps them blend into their surroundings, making them less visible to predators and potential threats.

In terms of their environment, harbor seals prefer coastal areas with easy access to both land and water. They can be spotted lounging on rocky shores, sandy beaches, or even floating on ice floes. These seals are adaptable and can inhabit a range of habitats, from cold, subarctic waters to more temperate coastal regions. Their ability to thrive in diverse environments contributes to their widespread distribution along coastlines.

The diet of harbor seals is primarily composed of fish, such as herring, cod, and salmon, along with crustaceans like shrimp and crabs. They are opportunistic feeders, taking advantage of available prey species in their habitat. Harbor seals are skilled hunters, using their sharp teeth and agile swimming abilities to catch their food. They are known to consume large quantities of fish daily, sustaining their energy needs and supporting their active lifestyles.

Behaviorally, harbor seals are social animals, often congregating in groups called "rafts" while resting or molting. They communicate with each other using a variety of vocalizations, including barks, grunts, and growls. These sounds serve various purposes, from maintaining social bonds to signaling danger or aggression. Harbor seals are also known for their playful nature, engaging in activities like swimming, diving, and exploring their surroundings.

Harbor seals have a fascinating life cycle, with females giving birth to their pups on land or ice floes. They care for their young, nursing them until they are ready to swim and hunt on their own. Molting is a regular part of a harbor seal's life, during which they shed their old fur and grow a new coat to maintain insulation and protection. These seals have excellent underwater vision, allowing them to spot prey and navigate effectively in murky waters. They are also capable of diving to impressive depths, reaching depths of 300 feet or more in search of food.

The presence of harbor seals in marine ecosystems plays a crucial role, contributing to nutrient cycling and supporting biodiversity through their interactions with prey species. Their adaptability, social behaviors, and ecological importance make them a fascinating species to study and appreciate in the coastal regions they call home.

Harp Seal

The harp seal, also known as the saddleback seal due to the distinctive markings on its back resembling a harp or saddle, is a fascinating marine mammal found in the icy waters of the Arctic and North Atlantic. Its physical appearance is characterized by a sleek, streamlined body with a silver-gray fur coat marked by dark patches on its back, giving it a striking and unique appearance. These seals have large, dark eyes that contribute to their adorable and expressive look.

Harp seals inhabit a wide range of environments, from the frigid Arctic waters to the subarctic regions of the North Atlantic. They are commonly found on pack ice and ice floes, where they haul out to rest, give birth, and nurse their young. These icy habitats provide them with essential shelter and access to their primary food sources.

The diet of harp seals mainly consists of fish, such as cod, herring, and capelin, as well as crustaceans like shrimp and krill. They are skilled hunters, using their sharp teeth and agile swimming abilities to catch their prey. Harp seals are opportunistic feeders, adapting their diet based on seasonal availability and migration patterns of their prey species.

Behaviorally, harp seals exhibit both solitary and social behaviors depending on the time of year. During the breeding season, they gather in large groups on pack ice, forming colonies where they give birth and mate. These colonies can consist of thousands of individuals, creating a bustling and lively atmosphere. Outside of the breeding season, harp seals may be more solitary, dispersing across their range to forage for food.

One interesting aspect of harp seals is their remarkable diving ability. They can dive to depths of over 1,000 feet and remain submerged for extended periods, allowing them to access deeper water where their prey may be abundant. This diving prowess is essential for their survival in the challenging Arctic environment.

Another fascinating fact about harp seals is their unique molting process. Unlike many other seal species, harp seals undergo a complete molt each year, shedding their old fur and growing a new coat. This molting cycle helps them maintain insulation and protect their skin from the harsh elements of their icy habitat.

Harp seals are also known for their vocalizations, producing a range of sounds including barks, grunts, and whistles. These vocalizations serve various purposes, from communication within colonies to expressing agitation or distress. They are highly social animals, engaging in playful behaviors such as sliding on ice and interacting with other seals.

One of the most iconic events in the life of a harp seal is the annual "whitecoat" stage of their pups. When born, harp seal pups have a pristine white fur coat, earning them the name "whitecoats." These adorable pups are nursed by their mothers until they are weaned and ready to venture into the water on their own.

In conclusion, the harp seal is a captivating marine mammal with a distinctive appearance, adaptive behaviors, and unique adaptations for survival in the Arctic and North Atlantic. Its role in these ecosystems as a predator and contributor to biodiversity underscores the importance of protecting and conserving these icy habitats.

Hooded Seal

The hooded seal is a remarkable marine mammal found in the cold waters of the North Atlantic and Arctic oceans. Its physical appearance is characterized by a large inflatable bladder, or "hood," located on the male seal's forehead. This unique feature is used during mating displays and can be inflated to an impressive size, creating a distinctive and striking appearance. The hooded seal's body is robust and cylindrical, with a dark gray to brownish-gray fur coat and lighter patches on its belly.

These seals primarily inhabit icy environments, including pack ice, ice floes, and coastal areas near Greenland, Iceland, and Canada. They are well adapted to the harsh conditions of the Arctic, using their thick blubber layer and dense fur for insulation against the cold. Hooded seals are known for their migratory patterns, traveling long distances between breeding and feeding grounds throughout the year.

The diet of hooded seals consists mainly of fish, such as capelin, herring, and cod, as well as squid and crustaceans. They are skilled hunters, using their sharp teeth and agile swimming abilities to catch prey. Hooded seals are opportunistic feeders, adjusting their diet based on the availability of food in their environment and seasonal changes in prey abundance.

Behaviorally, hooded seals exhibit a range of social behaviors depending on the time of year. During the breeding season, males gather in large groups called "leks" on pack ice, where they compete for mating opportunities with females. The inflatable hood is used as a display during these competitive gatherings, with males showcasing their size and dominance to attract mates.

One interesting aspect of hooded seals is their unique vocalizations. They produce a variety of sounds, including grunts, growls, and bell-like calls, which are used for communication within colonies and during mating displays. These vocalizations play a crucial role in social interactions and reproductive behaviors among hooded seals.

Another fascinating fact about hooded seals is their reproductive strategy. Females give birth to a single pup each year, typically on pack ice or ice floes. The pups are born with a soft, white fur coat that provides camouflage against the snowy backdrop of their environment. After weaning, the pups are left to fend for themselves and learn essential survival skills, such as swimming and hunting.

Hooded seals are also known for their impressive diving abilities. They can dive to depths of over 1,000 feet and remain submerged for extended periods, allowing them to access deep-water prey species and evade predators. This diving prowess is essential for their survival and successful foraging in the challenging Arctic environment.

In conclusion, the hooded seal is a fascinating marine mammal with unique physical features, adaptive behaviors, and specialized adaptations for life in icy waters. Its role in Arctic ecosystems as a predator and contributor to biodiversity highlights the importance of conservation efforts to protect these remarkable animals and their habitats.

Fur Seal

The fur seal is a captivating marine mammal found in various coastal regions around the world, including the Pacific and Southern oceans. Its physical appearance is characterized by a thick fur coat, which varies in color from light brown to dark gray, depending on the species. The fur seal's fur is incredibly dense and waterproof, providing excellent insulation against the cold waters in which they live. They have streamlined bodies with long flippers that allow them to move gracefully both in the water and on land.

These seals primarily inhabit rocky shorelines, beaches, and islands, where they haul out to rest, breed, and molt. They are well adapted to coastal environments, using rocky crevices and caves for shelter and protection. The specific habitats of fur seals can vary based on the species, with some preferring subantarctic islands and others inhabiting more temperate coastal areas.

The diet of fur seals consists mainly of fish, squid, and crustaceans, although their exact prey species can vary depending on their location and available food sources. They are skilled hunters, using their sharp teeth and agile swimming abilities to catch their prey. Fur seals are opportunistic feeders, adjusting their diet based on seasonal changes in prey abundance and migration patterns.

Behaviorally, fur seals are social animals, often forming colonies or groups called "rookeries" during the breeding season. These rookeries can be quite large, with thousands of individuals congregating to mate, give birth, and care for their young. Male fur seals are known for their territorial displays and vocalizations, which they use to establish dominance and attract mates.

One interesting aspect of fur seals is their unique reproductive strategies. Females typically give birth to a single pup each year, which they nurse and care for until the pup is weaned and independent. The bonding between mother and pup is strong, with females recognizing their offspring's vocalizations and scent among the crowded rookeries.

Fur seals are known for their agility and speed in the water, where they can swim with remarkable grace and precision. They use their powerful flippers to navigate through the ocean currents and pursue fast-moving prey. Their ability to dive to significant depths also contributes to their success as marine predators.

Another fascinating fact about fur seals is their annual molting process. During molting, fur seals shed their old fur coat and grow a new one, which helps them maintain insulation and protection against the elements. This molting cycle is essential for their survival and overall health, ensuring that they have a fresh and functional fur coat for the coming year.

In conclusion, fur seals are charismatic marine mammals with unique physical adaptations, social behaviors, and ecological roles in coastal ecosystems. Their fur coats, agile swimming abilities, and reproductive strategies make them fascinating subjects for study and observation, highlighting the diversity and complexity of life in the world's oceans.

Ring Seal

The ringed seal is a captivating marine mammal found in the Arctic and subarctic regions of the Northern Hemisphere. Its physical appearance is distinguished by distinctive light-colored rings or spots on its dark gray to brown fur coat, hence its name. These seals have compact, round bodies with short flippers and a small head, perfectly adapted for navigating through icy waters and hunting for food.

Ringed seals primarily inhabit polar and subpolar environments, including pack ice, ice floes, and coastal areas with access to open water. They are well adapted to the cold and harsh conditions of the Arctic, using their thick layer of blubber and dense fur for insulation against the frigid temperatures. These seals are highly specialized for life in icy habitats, where they can find refuge and safety from predators.

The diet of ringed seals consists mainly of fish, such as Arctic cod, herring, and salmon, as well as crustaceans like shrimp and krill. They are opportunistic feeders, using their sharp teeth and agile swimming abilities to catch their prey. Ringed seals are also known to create breathing holes in the ice, allowing them to access underwater prey and breathe while remaining hidden from predators like polar bears.

Behaviorally, ringed seals are solitary animals, spending much of their time alone or in small family groups. They are skilled divers, capable of diving to depths of over 300 feet and remaining submerged for extended periods. This diving ability is essential for their survival and foraging success in the icy waters of the Arctic.

One interesting aspect of ringed seals is their unique reproductive behavior. Females give birth to their pups in snow caves or on ice floes, where they create safe and sheltered environments for their young. The pups are born with a soft, white fur coat that provides camouflage against the snowy landscape. Female ringed seals are attentive mothers, nursing and caring for their pups until they are weaned and independent.

Ringed seals are also known for their vocalizations, which they use for communication and navigation in their icy environment. Their calls range from low grunts and growls to higher-pitched whistles and chirps, serving various purposes such as maintaining contact with other seals and identifying potential threats.

Another fascinating fact about ringed seals is their ability to create and maintain breathing holes in the ice using their strong claws and teeth. These breathing holes are crucial for their survival, allowing them to access air while swimming and diving in the icy waters of the Arctic.

In conclusion, the ringed seal is a remarkable marine mammal with unique physical adaptations, behaviors, and ecological significance in Arctic ecosystems. Its role as a predator, prey species, and contributor to biodiversity highlights the interconnectedness of life in polar regions and the importance of conservation efforts to protect these fascinating animals and their habitats.

Walrus

The walrus is a majestic marine mammal known for its impressive size, long tusks, and whiskered face. These large-bodied creatures can weigh up to several thousand pounds and have a distinctive appearance with their wrinkled skin, thick blubber layer, and prominent tusks. The tusks, which are elongated canine teeth, can grow to be several feet long and are used for a variety of purposes, including digging for food, hauling out of water onto ice, and establishing dominance within their social groups.

Walruses inhabit Arctic and subarctic regions, including the Arctic Ocean and the coasts of the North Atlantic and North Pacific. They are well adapted to cold environments, with specialized features such as a thick layer of blubber for insulation and dense fur that helps regulate body temperature. Walruses are often found near sea ice, ice floes, and coastal areas with shallow waters where they can access their preferred food sources.

The diet of walruses primarily consists of benthic invertebrates, such as clams, mussels, and other shellfish, which they locate by using their sensitive whiskers to detect vibrations in the seabed. They are skilled divers, capable of descending to depths of over 300 feet in search of food. Walruses use their muscular bodies and strong flippers to forage on the seafloor and extract prey from the sediment.

Behaviorally, walruses are highly social animals, often forming large herds or "haulouts" on ice floes or coastal areas. These gatherings serve various purposes, including mating, birthing, and resting. Male walruses, known as bulls, can be quite territorial during the breeding season, using their tusks and vocalizations to establish dominance and attract females.

One interesting aspect of walruses is their unique vocalizations, which include bell-like calls, grunts, and roars. These vocalizations are used for communication within their social groups, signaling danger, and maintaining social bonds. Walruses also use body language, such as posturing and head movements, to convey messages and establish hierarchies within their herds.

Another fascinating fact about walruses is their extensive use of sea ice and ice floes as platforms for resting, mating, and caring for their young. These ice habitats provide walruses with essential resources, including access to food, protection from predators like polar bears, and opportunities for social interactions.

Walruses are known for their playful behavior, often engaging in activities such as swimming, diving, and splashing in the water. They are curious animals and may investigate objects or boats that enter their territory, displaying a mix of caution and curiosity.

In conclusion, the walrus is a remarkable marine mammal with unique physical characteristics, behaviors, and ecological importance in Arctic and subarctic ecosystems. Its adaptations for life in cold environments, specialized feeding habits, and social behaviors make it a fascinating subject for study and observation, highlighting the diversity and complexity of life in polar regions.

Sea Lion

The sea lion is a captivating marine mammal known for its sleek, streamlined body, external ear flaps, and powerful flippers. These charismatic animals belong to the family Otariidae and are closely related to seals, but they can be distinguished by their visible ear flaps and ability to rotate their hind flippers forward, allowing them to walk on land more easily. Sea lions come in various sizes, with males typically larger than females and sporting a prominent mane of thick fur around their necks.

Sea lions inhabit coastal regions and islands around the world, including the North Pacific, North Atlantic, and Southern Hemisphere. They prefer rocky shorelines, sandy beaches, and coastal caves where they can haul out to rest, bask in the sun, and socialize with other members of their colony. These marine mammals are well adapted to life both in the water and on land, using their streamlined bodies and strong flippers for efficient swimming and maneuvering on land.

The diet of sea lions primarily consists of fish, squid, and crustaceans, although their specific prey species can vary depending on their location and seasonal availability. They are skilled hunters, using their sharp teeth and agile swimming abilities to catch their prey. Sea lions are also opportunistic feeders, taking advantage of a wide range of food sources in their marine environments.

Behaviorally, sea lions are highly social animals, often forming large colonies or groups called "rookeries" during the breeding season. These rookeries can consist of hundreds to thousands of individuals, creating a bustling and lively atmosphere. Sea lions communicate with each other using a variety of vocalizations, including barks, grunts, and growls, which serve various purposes such as maintaining social bonds and signaling danger.

One interesting aspect of sea lions is their agility and acrobatic skills in the water. They are excellent swimmers, capable of diving to significant depths and performing impressive underwater maneuvers. Sea lions use their flippers and streamlined bodies to glide through the water with speed and precision, making them efficient predators in their marine habitats.

Another fascinating fact about sea lions is their ability to communicate and navigate using sound. They produce a range of vocalizations that can travel long distances underwater, allowing them to locate prey, communicate with other sea lions, and navigate through their environment. These vocalizations play a crucial role in social interactions and survival strategies among sea lions.

Sea lions are also known for their intelligence and problem-solving abilities. They have been observed using tools, such as rocks or sticks, to aid in foraging or accessing hard-to-reach prey. This cognitive flexibility and adaptability contribute to their success as marine predators.

In conclusion, the sea lion is a captivating marine mammal with unique physical adaptations, social behaviors, and ecological importance in coastal ecosystems. Its role as a predator, social animal, and contributor to biodiversity highlights the interconnectedness of life in our oceans and the importance of conservation efforts to protect these fascinating animals and their habitats.

Sharks

Hammerhead Shark

The hammerhead shark is a fascinating marine predator known for its distinctive and unique appearance. Its most recognizable feature is its flattened, T-shaped head, which resembles a hammer or an anvil. This unusual head shape is thought to provide the shark with enhanced sensory capabilities, including a wider field of vision and improved maneuverability in the water. Hammerhead sharks come in various sizes, with some species reaching lengths of up to 20 feet or more.

These sharks inhabit warm coastal waters around the world, including tropical and subtropical regions of the Atlantic, Pacific, and Indian oceans. They prefer habitats with ample prey availability, such as coral reefs, continental shelves, and oceanic islands. Hammerhead sharks are highly migratory and can travel long distances in search of food, mating grounds, and suitable environments for breeding.

The diet of hammerhead sharks primarily consists of fish, squid, octopus, and crustaceans. They are skilled hunters, using their keen senses, including electroreception, to detect the electrical signals emitted by their prey. Hammerhead sharks have specialized ampullae of Lorenzini, sensory organs located in their heads, which help them locate prey hidden in the sand or camouflaged among rocks. They use their sharp, serrated teeth to capture and consume their prey efficiently.

Behaviorally, hammerhead sharks are solitary hunters, although they may form schools or groups during migration or breeding seasons. They are efficient swimmers, capable of rapid bursts of speed when pursuing prey. Hammerhead sharks also exhibit a unique behavior known as "spy-hopping," where they lift their heads above the water's surface to survey their surroundings or investigate potential prey items.

One interesting aspect of hammerhead sharks is their social structure and communication. While they are primarily solitary hunters, hammerhead sharks may form temporary aggregations or schools for feeding or mating purposes. They communicate using body language, posturing, and chemical signals, such as pheromones, to convey information about their intentions and establish social hierarchies.

Another fascinating fact about hammerhead sharks is their reproductive biology. They are viviparous, meaning they give birth to live young rather than laying eggs. Female hammerhead sharks carry their developing embryos in utero, nourishing them through a placental connection until they are ready to be born. Hammerhead shark pups are born fully formed and capable of swimming and hunting shortly after birth.

Hammerhead sharks have a unique swimming style characterized by their sweeping, undulating motions. This movement allows them to navigate through the water with agility and precision, making them efficient hunters in their marine environments. Their streamlined bodies and hydrodynamic shapes contribute to their speed and maneuverability.

In conclusion, the hammerhead shark is a fascinating and iconic marine predator with specialized physical adaptations, behaviors, and ecological roles in ocean ecosystems. Its distinctive head shape, sensory capabilities, hunting strategies, and reproductive biology make it a subject of interest and study among scientists and marine enthusiasts. Understanding and conserving these magnificent creatures are essential for maintaining the health and balance of marine ecosystems worldwide.

Great White Shark

The great white shark is a formidable apex predator known for its impressive size, sleek torpedo-shaped body, and rows of razor-sharp teeth. These sharks have a distinctively pointed snout and a large, powerful tail that propels them through the water with incredible speed and agility. Great white sharks are typically gray or blue-gray in color on their upper bodies, blending in with the ocean depths, while their bellies are lighter in color for camouflage from prey below.

These sharks are found in coastal waters around the world, particularly in temperate and tropical regions such as the coastal waters of Australia, South Africa, California, and the Mediterranean Sea. They prefer areas with abundant marine life, including seal colonies, fish-rich waters, and areas with large concentrations of migratory marine species. Great white sharks are highly migratory and can travel long distances in search of food and suitable breeding grounds.

The diet of great white sharks primarily consists of marine mammals such as seals, sea lions, and dolphins, as well as large fish such as tuna and rays. They are apex predators, meaning they are at the top of the marine food chain and play a crucial role in maintaining the balance of marine ecosystems. Great white sharks use their keen senses, including smell, sight, and electroreception, to locate and ambush their prey with precision.

Behaviorally, great white sharks are solitary hunters, although they may form loose aggregations or gather in areas with abundant food sources. They are opportunistic feeders, capable of scavenging on carrion and utilizing a wide range of prey items depending on availability. Great white sharks are known for their powerful and agile swimming abilities, allowing them to breach the water's surface in spectacular displays while hunting or investigating their surroundings.

One interesting aspect of great white sharks is their social behavior and communication. While they are primarily solitary hunters, great white sharks may engage in social interactions such as mating rituals, courtship displays, and aggressive encounters with other sharks. They communicate using body language, posturing, and chemical signals, conveying information about their intentions, dominance, and territorial boundaries.

Another fascinating fact about great white sharks is their reproductive biology. They are ovoviviparous, meaning they give birth to live young that develop inside eggs within the mother's body. Female great white sharks carry their developing embryos for up to 12 months before giving birth to a litter of pups. The pups are born fully formed and independent, capable of swimming and hunting shortly after birth.

Great white sharks have a unique hunting strategy known as "ambush predation," where they stalk and surprise their prey from below, launching upward with incredible speed and force to deliver a devastating bite. Their serrated teeth are designed for cutting and tearing flesh, allowing them to consume large prey items efficiently.

In conclusion, the great white shark is a fascinating and iconic marine predator with specialized physical adaptations, behaviors, and ecological roles in ocean ecosystems. Its role as an apex predator, hunting strategies, reproductive biology, and social behaviors make it a subject of interest and study among scientists and marine enthusiasts. Protecting and conserving these magnificent creatures is essential for maintaining the health and balance of marine ecosystems worldwide.

Nurse Shark

The nurse shark is a unique and fascinating species of shark known for its distinctive physical characteristics and behaviors. These sharks have a robust, cylindrical body with a broad, rounded snout and small eyes. They are typically gray or brown in color with a mottled pattern that provides camouflage against sandy or rocky seabeds. Nurse sharks have two dorsal fins, one near the tail and one closer to the head, and lack the sharp, pointed teeth typical of many other shark species.

Nurse sharks are primarily found in warm tropical and subtropical waters, particularly along the Atlantic coast of the Americas, from the southern United States to Brazil, as well as in parts of the Caribbean and Gulf of Mexico. They prefer shallow coastal areas such as coral reefs, mangrove forests, and sandy flats, where they can find ample food and shelter. Nurse sharks are bottom-dwellers, spending much of their time resting or hiding in caves, crevices, and under ledges during the day.

The diet of nurse sharks consists mainly of bottom-dwelling prey such as crustaceans, mollusks, small fish, and occasionally sea urchins and starfish. They are nocturnal feeders, using their powerful jaws and flat, crushing teeth to crush and consume their prey. Nurse sharks are opportunistic feeders, scavenging on carrion and feeding on a wide range of prey items depending on availability.

Behaviorally, nurse sharks are generally docile and non-aggressive toward humans, although they can become defensive if provoked or threatened. They are slow-moving and prefer to glide gracefully along the ocean floor, using their pectoral fins to maneuver and their tail to propel themselves through the water. Nurse sharks are known for their ability to "suction" food into their mouths by creating negative pressure with their gill covers, allowing them to feed efficiently on bottom-dwelling prey.

One interesting aspect of nurse sharks is their reproductive biology. They are ovoviviparous, meaning they give birth to live young that develop inside eggs within the mother's body. Female nurse sharks carry their developing embryos for up to six months before giving birth to a litter of pups, typically ranging from 20 to 30 pups per litter. The pups are born fully formed and independent, capable of swimming and hunting shortly after birth.

Another fascinating fact about nurse sharks is their ability to rest or "sleep" on the ocean floor during the day. They often seek out sheltered areas such as caves or ledges where they can rest undisturbed, using their buccal pumping mechanism to maintain oxygen flow through their gills while stationary.

Nurse sharks are known for their long lifespan, with some individuals living up to 25 years or more in the wild. This longevity allows them to play an important role in marine ecosystems as predators and scavengers, contributing to the balance of reef communities and helping regulate populations of bottom-dwelling prey species.

In conclusion, the nurse shark is a fascinating and important species in marine ecosystems, with unique physical adaptations, behaviors, and ecological roles. Its gentle demeanor, feeding habits, reproductive biology, and long lifespan make it a subject of interest and study among scientists and marine enthusiasts. Protecting and conserving nurse shark populations is essential for maintaining the health and diversity of marine habitats worldwide.

Bull Shark

The bull shark is a powerful and versatile predator known for its robust build, blunt snout, and aggressive behavior. These sharks have a stout, muscular body with a rounded snout and small eyes, giving them a distinctive appearance. They are typically gray or brown in color on their upper bodies, blending in with their surroundings, while their undersides are lighter in color for camouflage from prey below. Bull sharks have a stocky build and can grow to impressive sizes, with females generally larger than males.

Bull sharks are found in a wide range of marine environments, including coastal waters, estuaries, rivers, and even freshwater lakes. They are highly adaptable and can tolerate a wide range of salinities, allowing them to venture far inland along river systems. Bull sharks are commonly found in warm tropical and subtropical regions worldwide, such as the Gulf of Mexico, Caribbean Sea, and coastal waters of Africa, Asia, and Australia.

The diet of bull sharks is diverse and includes a variety of prey species such as fish, rays, crustaceans, turtles, and even small mammals. They are opportunistic feeders and skilled hunters, using their powerful jaws, sharp teeth, and strong swimming abilities to catch and consume their prey. Bull sharks are known for their aggressive feeding behavior, often using their momentum and strength to overpower larger prey items.

Behaviorally, bull sharks are known for their bold and assertive nature, earning them a reputation as one of the most aggressive shark species. They are curious and opportunistic hunters, exploring their environment and testing potential prey items with their teeth and jaws. Bull sharks are also territorial, particularly during mating and breeding seasons, and may display aggressive behavior toward other sharks or intruders in their territory.

One interesting aspect of bull sharks is their ability to tolerate a wide range of salinities and environments. They are one of the few shark species capable of entering freshwater habitats, such as rivers and lakes, where they can hunt for prey and even give birth to their young. This adaptability allows bull sharks to thrive in diverse ecosystems and exploit a wide range of food sources.

Another fascinating fact about bull sharks is their reproductive biology. They are viviparous, meaning they give birth to live young rather than laying eggs. Female bull sharks carry their developing embryos in utero for up to 12 months before giving birth to a litter of pups, typically ranging from 1 to 13 pups per litter. The pups are born fully formed and independent, capable of swimming and hunting shortly after birth.

Bull sharks are known for their impressive swimming abilities and can achieve speeds of up to 25 miles per hour when pursuing prey or navigating through their environment. Their streamlined bodies and powerful tail allow them to move swiftly and efficiently through the water, making them formidable predators in marine and freshwater habitats alike.

In conclusion, the bull shark is a formidable and adaptable predator with unique physical adaptations, behaviors, and ecological roles. Its aggressive nature, diverse diet, ability to thrive in various environments, and reproductive strategies make it a subject of interest and study among scientists and marine enthusiasts. Understanding and conserving bull shark populations are essential for maintaining the health and balance of marine and freshwater ecosystems worldwide.

Tiger Shark

The tiger shark is a large and powerful marine predator known for its distinctive appearance and voracious appetite. These sharks have a robust, cylindrical body with a broad, blunt snout and small eyes. They are typically gray or bluish-gray in color on their upper bodies, with vertical stripes or patterns that resemble those of a tiger, giving them their name. Tiger sharks have serrated teeth and a powerful jaw, enabling them to consume a wide range of prey items.

Tiger sharks are found in warm and temperate waters around the world, including the Atlantic, Pacific, and Indian oceans. They prefer coastal areas, coral reefs, and offshore islands where they can find abundant food sources. Tiger sharks are highly adaptable and can inhabit a variety of marine environments, from shallow coastal waters to deep oceanic habitats.

The diet of tiger sharks is diverse and includes a wide range of prey species such as fish, rays, sea turtles, seabirds, and even marine mammals. They are opportunistic feeders and skilled hunters, using their keen senses, including smell, sight, and electroreception, to locate and capture their prey. Tiger sharks are known for their scavenging behavior and are often referred to as "garbage cans of the sea" due to their ability to consume almost anything, including human-made objects found in their environment.

Behaviorally, tiger sharks are solitary hunters and prefer to roam their territories in search of food. They are known for their bold and curious nature, often investigating potential prey items or objects in their environment with their mouths. Tiger sharks are also highly territorial and may display aggressive behavior toward other sharks or intruders in their territory.

One interesting aspect of tiger sharks is their reproductive biology. They are viviparous, meaning they give birth to live young rather than laying eggs. Female tiger sharks carry their developing embryos in utero for up to 16 months before giving birth to a litter of pups, typically ranging from 10 to 80 pups per litter. The pups are born fully formed and independent, capable of swimming and hunting shortly after birth.

Tiger sharks are known for their impressive swimming abilities and can cover long distances in search of food. They have a unique hunting strategy known as "bump and bite," where they use their snouts to bump into potential prey items to assess their edibility before delivering a powerful bite. This strategy allows them to conserve energy and avoid consuming undesirable or harmful prey.

Another fascinating fact about tiger sharks is their ability to consume a wide range of food items, including unusual and unexpected prey. They have been known to eat objects such as tires, license plates, and even clothing found in their environment. This adaptability and opportunistic feeding behavior contribute to their success as apex predators in marine ecosystems.

Tiger sharks play an important role in marine ecosystems as top predators, helping to regulate populations of prey species and maintain the balance of marine food webs. Their presence is indicative of healthy and diverse ocean habitats, making them a vital component of marine conservation efforts.

In conclusion, the tiger shark is a formidable and adaptable predator with unique physical adaptations, behaviors, and ecological roles. Its distinctive appearance, diverse diet, reproductive strategies, and hunting techniques make it a subject of interest and study among scientists and marine enthusiasts. Protecting and conserving tiger shark populations is essential for maintaining the health and resilience of marine ecosystems worldwide.

Thresher Shark

The thresher shark is a remarkable marine predator distinguished by its elongated upper lobe of the tail fin, which can be as long as the shark's body itself. This unique feature gives the thresher shark a distinct appearance and aids in its agile swimming capabilities. They have a streamlined body with a pointed snout and large eyes, perfectly adapted for hunting in the open ocean. Thresher sharks are typically gray or brown in color, with lighter undersides for camouflage against the sunlight filtering through the water.

These sharks are primarily found in temperate and tropical waters worldwide, including the Atlantic, Pacific, and Indian oceans. They prefer offshore habitats and deep-sea environments, where they can find abundant prey such as small fish, squid, and pelagic crustaceans. Thresher sharks are highly migratory and may travel long distances in search of food and suitable breeding grounds.

The diet of thresher sharks consists mainly of small fish, including herring, mackerel, and sardines, as well as squid and occasionally crustaceans. They are skilled hunters, using their elongated tail fin to stun and herd schools of fish before striking with their sharp teeth. Thresher sharks are known for their unique hunting technique called "tail slapping," where they use their long tail fin to slap the water's surface and stun prey.

Behaviorally, thresher sharks are solitary hunters and prefer to roam their territories in search of food. They are highly agile and capable swimmers, using their powerful tail fin and body movements to navigate through the water with speed and precision. Thresher sharks are known for their bursts of speed when pursuing prey, reaching speeds of up to 30 miles per hour.

One interesting aspect of thresher sharks is their reproductive biology. They are ovoviviparous, meaning they give birth to live young that develop inside eggs within the mother's body. Female thresher sharks carry their developing embryos for up to nine months before giving birth to a litter of pups, typically ranging from 2 to 4 pups per litter. The pups are born fully formed and independent, capable of swimming and hunting shortly after birth.

Thresher sharks have a unique anatomy that allows them to maintain buoyancy and stability in the water. Their large pectoral fins and elongated tail fin provide lift and propulsion, allowing them to glide effortlessly through the water column. This specialized anatomy contributes to their efficiency as predators in the open ocean.

Another fascinating fact about thresher sharks is their communication and social behavior. While they are primarily solitary hunters, thresher sharks may form loose aggregations or schools during migration or mating seasons. They communicate using body language, posturing, and chemical signals, conveying information about their intentions and establishing social hierarchies within their groups.

Thresher sharks play an important role in marine ecosystems as top predators, helping to regulate populations of prey species and maintain the balance of ocean food webs. Their presence is indicative of healthy and diverse ocean habitats, making them a vital component of marine conservation efforts.

In conclusion, the thresher shark is a remarkable and adaptable predator with unique physical adaptations, behaviors, and ecological roles. Its distinctive appearance, hunting techniques, reproductive strategies, and social behaviors make it a subject of interest and study among scientists and marine enthusiasts. Protecting and conserving thresher shark populations is essential for maintaining the health and resilience of marine ecosystems worldwide.

Blue Shark

The blue shark is a sleek and agile marine predator known for its striking blue coloration and streamlined body. These sharks have a long, slender body with a pointed snout, large eyes, and a distinctive blue-gray coloration on their upper bodies that fades to a lighter shade on their undersides. Their bodies are designed for efficient swimming, with a hydrodynamic shape and powerful tail fin that allows them to move gracefully through the water.

Blue sharks are primarily found in temperate and tropical waters worldwide, including the Atlantic, Pacific, and Indian oceans. They prefer offshore habitats and deep-sea environments, where they can find abundant prey such as squid, fish, and small pelagic crustaceans. Blue sharks are highly migratory and may travel long distances in search of food and suitable breeding grounds.

The diet of blue sharks consists mainly of small fish, including mackerel, herring, and sardines, as well as squid and occasionally crustaceans. They are opportunistic feeders and skilled hunters, using their keen senses, including smell, sight, and electroreception, to locate and capture their prey. Blue sharks are known for their ability to detect and follow scent trails in the water, allowing them to track down prey over long distances.

Behaviorally, blue sharks are active and agile swimmers, using their powerful tail fin and body movements to navigate through the water with speed and precision. They are capable of rapid bursts of speed when pursuing prey or avoiding predators, reaching speeds of up to 40 miles per hour. Blue sharks are also known for their curiosity and may investigate objects or boats in their environment with caution.

One interesting aspect of blue sharks is their reproductive biology. They are ovoviviparous, meaning they give birth to live young that develop inside eggs within the mother's body. Female blue sharks carry their developing embryos for up to nine months before giving birth to a litter of pups, typically ranging from 4 to 135 pups per litter. The pups are born fully formed and independent, capable of swimming and hunting shortly after birth.

Blue sharks have a unique anatomy that allows them to maintain buoyancy and stability in the water. Their large pectoral fins and streamlined body shape provide lift and maneuverability, allowing them to navigate through the water column with ease. This specialized anatomy contributes to their efficiency as predators in the open ocean.

Another fascinating fact about blue sharks is their migratory behavior and long-distance travel patterns. They are known to undertake extensive migrations, moving between feeding grounds and breeding areas over thousands of miles. Blue sharks may also form loose aggregations or schools during migration, providing opportunities for social interactions and mating.

Blue sharks play an important role in marine ecosystems as top predators, helping to regulate populations of prey species and maintain the balance of ocean food webs. Their presence is indicative of healthy and diverse ocean habitats, making them a vital component of marine conservation efforts.

In conclusion, the blue shark is a sleek and agile predator with unique physical adaptations, behaviors, and ecological roles. Its striking blue coloration, hunting techniques, reproductive strategies, and migratory behavior make it a subject of interest and study among scientists and marine enthusiasts. Protecting and conserving blue shark populations is essential for maintaining the health and resilience of marine ecosystems worldwide.

Mako Shark

The mako shark is a sleek and powerful marine predator known for its streamlined body, pointed snout, and impressive swimming abilities. These sharks have a torpedo-shaped body with a conical snout, large eyes, and a distinctively long and curved tail fin. Mako sharks are typically blue or metallic blue-gray in color on their upper bodies, blending in with the open ocean, while their undersides are lighter in color for camouflage from prey below.

Mako sharks inhabit warm and temperate waters worldwide, including the Atlantic, Pacific, and Indian oceans. They prefer offshore habitats and deep-sea environments, where they can find abundant prey such as fish, squid, and pelagic crustaceans. Mako sharks are highly migratory and may travel long distances in search of food and suitable breeding grounds.

The diet of mako sharks consists mainly of small to medium-sized fish, including mackerel, tuna, and swordfish, as well as squid and occasionally crustaceans. They are opportunistic feeders and skilled hunters, using their keen senses, including sight, smell, and electroreception, to locate and capture their prey. Mako sharks are known for their speed and agility in chasing down fast-moving prey, such as schools of fish or agile squid.

Behaviorally, mako sharks are active and swift swimmers, using their powerful tail fin and body movements to navigate through the water with speed and precision. They are capable of rapid bursts of speed when pursuing prey or avoiding predators, reaching speeds of up to 60 miles per hour. Mako sharks are also known for their acrobatic leaps out of the water, known as breaching, which they may use as a hunting technique or to escape threats.

One interesting aspect of mako sharks is their reproductive biology. They are ovoviviparous, meaning they give birth to live young that develop inside eggs within the mother's body. Female mako sharks carry their developing embryos for up to 15 months before giving birth to a litter of pups, typically ranging from 4 to 18 pups per litter. The pups are born fully formed and independent, capable of swimming and hunting shortly after birth.

Mako sharks have a unique anatomy that contributes to their efficiency as predators in the open ocean. Their streamlined body shape, powerful tail fin, and sharp teeth allow them to move swiftly through the water and capture prey with precision. They are also known for their keen senses, including excellent eyesight and a highly developed sense of smell, which help them locate and track down prey over long distances.

Another fascinating fact about mako sharks is their role as top predators in marine ecosystems. They help regulate populations of prey species and maintain the balance of ocean food webs. Mako sharks are also known to engage in social behaviors, such as mating rituals and courtship displays, during breeding seasons.

In conclusion, the mako shark is a remarkable and formidable predator with unique physical adaptations, behaviors, and ecological roles. Its sleek body, impressive swimming abilities, hunting techniques, reproductive strategies, and role as a top predator make it a subject of interest and study among scientists and marine enthusiasts. Protecting and conserving mako shark populations is essential for maintaining the health and resilience of marine ecosystems worldwide.

Megalodon

The Megalodon was an ancient and colossal marine predator that roamed the oceans millions of years ago. It is believed to have been one of the largest and most formidable predators to have ever existed, with a size estimated to be three times that of a modern great white shark. The physical appearance of the Megalodon is inferred from fossil evidence, including massive teeth measuring up to seven inches in length and a jaw structure capable of delivering devastating bites.

Megalodons inhabited warm and temperate waters worldwide during the Miocene and Pliocene epochs, from approximately 23 million to 3.6 million years ago. They preferred coastal and offshore habitats where prey was abundant, such as ancient seas and shallow continental shelves. Megalodons were apex predators, meaning they were at the top of the marine food chain and played a crucial role in shaping ancient marine ecosystems.

The diet of Megalodons consisted primarily of large marine mammals such as whales, dolphins, and seals, as well as other large fish and sea turtles. They were opportunistic feeders, using their massive size, sharp teeth, and powerful jaws to capture and consume their prey. Megalodons were likely skilled hunters, employing ambush tactics and powerful swimming abilities to pursue and overwhelm their targets.

Behaviorally, Megalodons were likely solitary hunters, although they may have gathered in areas with abundant prey or during mating seasons. They were highly adapted for marine life, with a streamlined body shape, large pectoral fins for maneuverability, and a robust tail for propulsion. Megalodons are thought to have been capable of fast bursts of speed when hunting, allowing them to surprise and capture agile prey.

One fascinating aspect of Megalodons is their extinction and evolutionary history. They disappeared from the fossil record around 3.6 million years ago, possibly due to changes in climate, ocean currents, or competition with other predators. The extinction of Megalodons had a significant impact on marine ecosystems, leading to shifts in predator-prey dynamics and the evolution of new marine species.

Another interesting fact about Megalodons is their size and power. They are estimated to have reached lengths of up to 60 feet or more, making them one of the largest predators to have ever lived. Their massive teeth, robust jaws, and formidable hunting abilities made them apex predators in ancient oceans.

Megalodons are a subject of fascination and study among paleontologists, marine biologists, and enthusiasts interested in prehistoric life. Fossil discoveries, scientific research, and computer simulations have provided valuable insights into the biology, behavior, and ecological role of these ancient giants. Understanding the life and times of Megalodons helps us better appreciate the diversity and complexity of ancient marine ecosystems.

In conclusion, Megalodons were massive and formidable predators that dominated ancient oceans millions of years ago. Their physical appearance, habitat preferences, diet, behavior, extinction, and evolutionary history make them a subject of enduring interest and curiosity. Studying Megalodons provides valuable insights into the dynamics of prehistoric marine ecosystems and the evolution of marine life on Earth.

Jellyfish

Lion's Mane Jellyfish

The lion's mane jellyfish is a mesmerizing marine creature known for its striking appearance and unique characteristics. It belongs to the phylum Cnidaria and is the largest known species of jellyfish, with some individuals reaching sizes of over six feet in diameter. The physical appearance of the lion's mane jellyfish is characterized by its bell-shaped body, long tentacles, and vibrant coloration, ranging from deep red to orange or yellow.

These jellyfish are found in cold and temperate waters throughout the world's oceans, particularly in the northern Atlantic and Pacific oceans. They prefer coastal areas and can often be found near shorelines, estuaries, and bays where nutrient-rich waters support abundant prey. Lion's mane jellyfish are also known to inhabit deeper oceanic habitats, where they drift with ocean currents in search of food.

The diet of lion's mane jellyfish primarily consists of small fish, plankton, crustaceans, and other jellyfish species. They are passive drifters, using their long, trailing tentacles equipped with stinging cells called nematocysts to capture and immobilize prey. Once caught, the tentacles transport the prey to the jellyfish's mouth located at the center of its bell-shaped body for consumption.

Behaviorally, lion's mane jellyfish are relatively slow-moving and rely on ocean currents for transportation and dispersal. They are non-aggressive towards humans and other creatures, but their tentacles can deliver a painful sting if touched. Lion's mane jellyfish are known for their ability to pulsate their bell-shaped bodies, propelling themselves through the water in a slow and graceful manner.

One interesting aspect of lion's mane jellyfish is their anatomy and structure. They have a complex network of tentacles that can extend for several feet, allowing them to capture prey over a wide area. The bell-shaped body of the jellyfish contains a central cavity with a mouth opening, surrounded by a ring of oral arms used for feeding and digestion.

Another fascinating fact about lion's mane jellyfish is their size variability. While some individuals can grow to enormous sizes, others may be much smaller, with bell diameters ranging from a few inches to several feet. This variability in size is influenced by factors such as age, environmental conditions, and availability of food.

Lion's mane jellyfish play an important role in marine ecosystems as both predators and prey. They help regulate populations of plankton and small marine organisms, contributing to the balance of ocean food webs. Their presence also provides food for various marine predators, including sea turtles, fish, and seabirds.

One intriguing aspect of lion's mane jellyfish is their life cycle and reproductive strategies. They undergo a complex life cycle that includes both sexual and asexual reproduction. During the breeding season, adult jellyfish release sperm and eggs into the water, where fertilization occurs. The fertilized eggs develop into free-swimming larvae called planulae, which eventually settle on the ocean floor and develop into polyps. These polyps then bud off and produce medusae, or jellyfish, completing the life cycle.

In conclusion, the lion's mane jellyfish is a captivating and important species in marine ecosystems. Its physical appearance, habitat preferences, diet, behavior, and life cycle make it a subject of interest and study among marine biologists and enthusiasts. Understanding and conserving lion's mane jellyfish populations are crucial for maintaining the health and balance of ocean ecosystems worldwide.

Moon Jellyfish

The moon jellyfish is a fascinating and delicate marine creature known for its translucent bell-shaped body and gentle drifting movements. It belongs to the phylum Cnidaria and is one of the most common species of jellyfish found in oceans worldwide. The physical appearance of the moon jellyfish is characterized by its round, saucer-like bell that can range in diameter from a few inches to several feet. The bell is typically transparent or pale in color, with four horseshoe-shaped gonads visible through the bell's surface.

Moon jellyfish are found in a wide range of marine environments, including coastal waters, estuaries, and open oceans. They are particularly abundant in temperate and tropical regions but can also thrive in colder waters. Moon jellyfish are well-adapted to a variety of habitats and can tolerate a wide range of salinity and temperature conditions.

The diet of moon jellyfish primarily consists of small planktonic organisms such as zooplankton, fish eggs, larvae, and small crustaceans. They are passive feeders, using their tentacles equipped with stinging cells called nematocysts to capture prey. Once caught, the prey is transported to the jellyfish's mouth located at the center of its bell-shaped body for ingestion and digestion.

Behaviorally, moon jellyfish are gentle drifters, relying on ocean currents and water movements to navigate and feed. They are non-aggressive and pose little threat to humans or other marine creatures. Moon jellyfish move by contracting and relaxing their bell-shaped bodies, pulsating in a rhythmic manner that propels them through the water with graceful movements.

One interesting aspect of moon jellyfish is their unique reproductive strategy. They undergo a complex life cycle that includes both sexual and asexual reproduction. During the breeding season, adult jellyfish release sperm and eggs into the water, where fertilization occurs externally. The fertilized eggs develop into larvae called planulae, which eventually settle on the ocean floor and develop into polyps. These polyps then bud off and produce medusae, or jellyfish, completing the life cycle.

Another fascinating fact about moon jellyfish is their bioluminescent capabilities. They can emit a soft glow or luminescence, particularly at night or in darker waters. This bioluminescence is thought to serve as a form of camouflage or communication, helping moon jellyfish avoid predators or attract mates.

Moon jellyfish are an important part of marine ecosystems as they serve as food for various marine predators, including sea turtles, fish, and seabirds. Their presence also helps regulate populations of planktonic organisms, contributing to the balance of ocean food webs.

One intriguing aspect of moon jellyfish is their ability to undergo a process called transdifferentiation. This means that certain cells within their bodies can transform into different cell types, allowing moon jellyfish to repair damaged tissues or regenerate lost body parts. This remarkable ability contributes to their resilience and survival in changing environmental conditions.

In conclusion, the moon jellyfish is a captivating and important species in marine ecosystems. Its physical appearance, habitat preferences, diet, behavior, and unique characteristics make it a subject of interest and study among marine biologists and enthusiasts. Understanding and conserving moon jellyfish populations are crucial for maintaining the health and balance of ocean ecosystems worldwide.

Cannonball Jellyfish

The cannonball jellyfish, also known as the cabbagehead jellyfish, is a distinctive marine creature with a unique appearance and ecological significance. It belongs to the phylum Cnidaria and is characterized by its round, dome-shaped bell and short, stubby tentacles. The bell of the cannonball jellyfish is typically smooth and lacks the long, trailing tentacles found in other jellyfish species.

These jellyfish are commonly found in coastal and estuarine waters along the Atlantic and Gulf coasts of North America, as well as in other temperate and tropical regions worldwide. They prefer shallow, brackish waters near shorelines, marshes, and mangrove forests, where nutrient-rich currents support abundant planktonic food sources.

The diet of cannonball jellyfish consists mainly of small planktonic organisms such as zooplankton, copepods, and larval fish. They are passive feeders, using their short tentacles equipped with stinging cells called nematocysts to capture prey. Cannonball jellyfish are not aggressive towards humans and pose little threat due to their mild sting.

Behaviorally, cannonball jellyfish are gentle drifters, relying on ocean currents and water movements to navigate and feed. They are non-aggressive and prefer to avoid confrontation with other marine creatures. Cannonball jellyfish move by pulsating their bell-shaped bodies in a slow and rhythmic manner, propelling themselves through the water with gentle movements.

One interesting aspect of cannonball jellyfish is their role in marine ecosystems as both predators and prey. They help regulate populations of planktonic organisms, contributing to the balance of ocean food webs. Cannonball jellyfish also serve as food for various marine predators, including sea turtles, fish, and seabirds.

Another fascinating fact about cannonball jellyfish is their reproductive biology. They undergo a complex life cycle that includes both sexual and asexual reproduction. During the breeding season, adult jellyfish release sperm and eggs into the water, where fertilization occurs externally. The fertilized eggs develop into larvae called planulae, which eventually settle on the ocean floor and develop into polyps. These polyps then bud off and produce medusae, or jellyfish, completing the life cycle.

Cannonball jellyfish are often used as a food source in some cultures and cuisines. In parts of Asia, they are considered a delicacy and are consumed in dishes such as salads, soups, and stir-fries. However, caution should be exercised when handling or consuming cannonball jellyfish, as their mild sting can cause discomfort in some individuals.

One intriguing aspect of cannonball jellyfish is their symbiotic relationship with certain marine species. They are known to host symbiotic algae called zooxanthellae within their tissues, which provide them with energy through photosynthesis. This symbiosis allows cannonball jellyfish to thrive in nutrient-poor environments and enhances their resilience to environmental changes.

In conclusion, the cannonball jellyfish is a unique and important species in marine ecosystems. Its physical appearance, habitat preferences, diet, behavior, and ecological roles make it a subject of interest and study among marine biologists and enthusiasts. Understanding and conserving cannonball jellyfish populations are crucial for maintaining the health and balance of coastal and estuarine ecosystems worldwide.

Barrel Jellyfish

The barrel jellyfish is a fascinating marine species known for its distinctive appearance and ecological significance. It belongs to the phylum Cnidaria and is characterized by its large, bell-shaped body and long, frilly oral arms. The bell of the barrel jellyfish can reach sizes of up to three feet in diameter, making it one of the largest jellyfish species in the world. The body is usually translucent or pale white in color, with a dome-shaped structure that gives it a barrel-like appearance.

Barrel jellyfish are commonly found in coastal and offshore waters throughout the world's oceans, including the Atlantic, Pacific, and Indian oceans. They prefer temperate and subtropical regions but can also be found in colder waters during certain times of the year. Barrel jellyfish are often seen in estuaries, bays, and coastal areas where nutrient-rich currents support abundant planktonic food sources.

The diet of barrel jellyfish consists mainly of small planktonic organisms such as zooplankton, copepods, and small fish larvae. They are passive feeders, using their long oral arms equipped with stinging cells called nematocysts to capture prey. Barrel jellyfish are not aggressive towards humans and pose little threat due to their mild sting.

Behaviorally, barrel jellyfish are gentle drifters, relying on ocean currents and water movements to navigate and feed. They are non-aggressive and prefer to avoid confrontation with other marine creatures. Barrel jellyfish move by pulsating their bell-shaped bodies in a slow and rhythmic manner, propelling themselves through the water with graceful movements.

One interesting aspect of barrel jellyfish is their symbiotic relationship with certain marine species. They are known to host symbiotic algae called zooxanthellae within their tissues, which provide them with energy through photosynthesis. This symbiosis allows barrel jellyfish to thrive in nutrient-poor environments and enhances their resilience to environmental changes.

Another fascinating fact about barrel jellyfish is their reproductive biology. They undergo a complex life cycle that includes both sexual and asexual reproduction. During the breeding season, adult jellyfish release sperm and eggs into the water, where fertilization occurs externally. The fertilized eggs develop into larvae called planulae, which eventually settle on the ocean floor and develop into polyps. These polyps then bud off and produce medusae, or jellyfish, completing the life cycle.

Barrel jellyfish are often encountered by divers and snorkelers in coastal areas, where they can form large aggregations or blooms during certain times of the year. These blooms are usually harmless to humans but can be a spectacular sight to witness underwater.

In conclusion, the barrel jellyfish is a unique and important species in marine ecosystems. Its physical appearance, habitat preferences, diet, behavior, and ecological roles make it a subject of interest and study among marine biologists and enthusiasts. Understanding and conserving barrel jellyfish populations are crucial for maintaining the health and balance of coastal and offshore ecosystems worldwide.

Box Jellyfish

The box jellyfish is a remarkable marine creature known for its unique physical appearance and potent venom. It belongs to the phylum Cnidaria and is characterized by its cube-shaped bell and long, trailing tentacles armed with venomous stingers called nematocysts. The box jellyfish's bell can grow up to ten inches in diameter, and its tentacles can extend to several feet in length, making it a formidable predator in the ocean.

These jellyfish are typically found in warm coastal waters of the Indo-Pacific region, including the waters of Australia, Southeast Asia, and the Pacific Islands. They prefer shallow, tropical environments with ample food sources and suitable breeding grounds. Box jellyfish are most commonly encountered during the warmer months when water temperatures are conducive to their activity.

The diet of box jellyfish consists mainly of small fish, crustaceans, and other jellyfish species. They are active hunters, using their long tentacles and potent venom to immobilize and capture prey. Box jellyfish are known for their fast and agile swimming abilities, allowing them to pursue and catch fast-moving prey with precision.

Behaviorally, box jellyfish are highly adapted predators with specialized sensory structures that allow them to detect and respond to environmental cues. They have complex eyes with a 360-degree field of vision, enabling them to detect light and movement in their surroundings. Box jellyfish are also capable of coordinated movements and navigation, allowing them to actively seek out prey and avoid obstacles in the water.

One interesting aspect of box jellyfish is their potent venom, which contains toxins that can cause severe pain, tissue damage, and in some cases, even death in humans. The venomous stingers on their tentacles can deliver a powerful sting upon contact, making them one of the most dangerous jellyfish species to encounter in the ocean. It is important for swimmers and beachgoers in areas where box jellyfish are present to exercise caution and follow safety guidelines to avoid stings.

Another fascinating fact about box jellyfish is their reproductive biology. They undergo a complex life cycle that includes both sexual and asexual reproduction. During the breeding season, adult jellyfish release sperm and eggs into the water, where fertilization occurs externally. The fertilized eggs develop into larvae called planulae, which eventually settle on the ocean floor and develop into polyps. These polyps then bud off and produce medusae, or jellyfish, completing the life cycle.

Box jellyfish are known for their unique swimming behavior, which involves rhythmic contractions of their bell-shaped bodies and coordinated movements of their tentacles. This swimming style allows them to move swiftly through the water and capture prey with precision. Box jellyfish are also capable of changing their swimming direction and speed in response to environmental stimuli, such as changes in water currents or the presence of potential threats.

In conclusion, the box jellyfish is a fascinating and formidable predator in marine ecosystems. Its physical appearance, habitat preferences, diet, behavior, and potent venom make it a subject of interest and study among marine biologists and enthusiasts. Understanding and respecting the presence of box jellyfish in coastal waters are essential for promoting safety and conservation efforts to protect both humans and marine life.

Portuguese Man of War

The Portuguese man o' war is a striking marine organism known for its unique appearance and potent sting. It is not a true jellyfish but rather a colonial organism called a siphonophore, consisting of a colony of specialized polyps working together as one entity. The physical appearance of the Portuguese man o' war is characterized by its gas-filled float, which resembles a translucent balloon or sail, often blue or purple in color, with long tentacles extending below the water's surface.

These organisms are commonly found in warm and tropical waters worldwide, including the Atlantic, Pacific, and Indian oceans. They prefer open ocean environments with ample sunlight and nutrient-rich waters, where they can drift with ocean currents. Portuguese man o' war are often seen floating on the surface of the water, propelled by the wind and currents.

The diet of Portuguese man o' war primarily consists of small fish, crustaceans, and other planktonic organisms. They are passive predators, using their long tentacles armed with venomous nematocysts to capture and immobilize prey. The tentacles can extend several feet below the surface, allowing the Portuguese man o' war to catch prey in the water column.

Behaviorally, Portuguese man o' war are highly adapted to life in the open ocean. They rely on ocean currents and wind for movement, drifting with the water's surface in search of food and suitable breeding grounds. Portuguese man o' war are not capable of independent swimming but instead use their float and tentacles to navigate and capture prey.

One interesting aspect of Portuguese man o' war is their reproductive biology. They reproduce both sexually and asexually, with colonies releasing sperm and eggs into the water for fertilization. The fertilized eggs develop into larvae, which eventually settle and grow into new colonies of Portuguese man o' war. A single colony can release thousands of eggs and larvae, contributing to their widespread distribution in oceanic waters.

Another fascinating fact about Portuguese man o' war is their powerful sting. While their tentacles may look delicate, they are equipped with venomous nematocysts that can cause intense pain and irritation in humans and other animals. Contact with the tentacles can result in skin irritation, swelling, and in some cases, allergic reactions or systemic effects. It is important to exercise caution and avoid touching Portuguese man o' war if encountered in the water or on the beach.

Portuguese man o' war are often mistaken for jellyfish due to their similar appearance, but they are distinct organisms with unique characteristics. Unlike jellyfish, Portuguese man o' war are composed of multiple specialized polyps working together as a colony, each with specific functions such as feeding, reproduction, and defense.

In conclusion, the Portuguese man o' war is a fascinating and enigmatic marine organism with a unique appearance, ecological role, and powerful sting. Its physical appearance, habitat preferences, diet, behavior, and reproductive strategies make it a subject of interest and study among marine biologists and enthusiasts. Understanding and respecting the presence of Portuguese man o' war in oceanic waters are essential for promoting safety and conservation efforts to protect both humans and marine life.

Reef Fish

Clownfish

The clownfish is a vibrant and fascinating marine fish known for its distinctive appearance and unique symbiotic relationship with sea anemones. These small fish belong to the family Pomacentridae and are characterized by their bright colors, including orange, black, white, and yellow stripes. The physical appearance of clownfish includes a compressed and elongated body, rounded fins, and a prominent mouth with sharp teeth.

Clownfish are typically found in warm and tropical waters of the Indian and Pacific oceans, particularly in coral reefs and lagoons. They prefer shallow waters with ample hiding places such as coral crevices or sea anemones. Clownfish have a mutualistic relationship with certain species of sea anemones, where they seek protection and shelter among the anemone's tentacles in exchange for food and waste.

The diet of clownfish primarily consists of small crustaceans, zooplankton, algae, and leftover food from the sea anemone's meals. They are omnivorous feeders, consuming a variety of prey items and algae to meet their nutritional needs. Clownfish are also known to scavenge for food and may feed on small invertebrates and organic matter found in their habitat.

Behaviorally, clownfish are social and territorial fish that form hierarchical groups within their coral reef habitats. They establish and defend territories around their host sea anemone, using aggressive displays and vocalizations to ward off intruders and protect their nesting sites. Clownfish exhibit complex social behaviors, including courtship rituals, cooperative breeding, and communication through visual cues and chemical signals.

One interesting aspect of clownfish is their unique symbiotic relationship with sea anemones. They are immune to the venomous tentacles of the anemone and use them as protection from predators. In return, clownfish provide the sea anemone with nutrients and protection from potential threats. This mutually beneficial partnership benefits both species and enhances their survival in coral reef ecosystems.

Another fascinating fact about clownfish is their ability to change sex. In clownfish social groups, there is a dominant female and a breeding male, with subordinate males and non-breeding females forming a hierarchy. If the dominant female dies or is removed, the breeding male will transition into a female, and the next-highest-ranking male will become the new breeding male. This sex change ensures the continuity of breeding within the group.

Clownfish are known for their unique swimming patterns and behaviors. They move in a zigzagging motion and often dart in and out of coral crevices or anemone tentacles, making them agile and difficult for predators to catch. Clownfish also exhibit territorial behavior, defending their nesting sites and food resources within their coral reef habitat.

In conclusion, the clownfish is a colorful and charismatic species with a fascinating ecology and behavior. Its physical appearance, habitat preferences, diet, behavior, and symbiotic relationship with sea anemones make it a subject of interest and study among marine biologists and enthusiasts. Understanding and conserving clownfish populations are crucial for maintaining the health and diversity of coral reef ecosystems worldwide.

Yellow Tang

The yellow tang is a vibrant and popular marine fish known for its bright yellow coloration and graceful swimming behavior. This species, scientifically known as Zebrasoma flavescens, belongs to the family Acanthuridae and is native to the coral reefs of the Indo-Pacific region. Its physical appearance includes an oval-shaped body, small mouth with fine teeth, and a continuous dorsal fin that runs along its back.

Yellow tangs are commonly found in shallow, tropical waters with coral reef habitats, including areas around Hawaii, the Red Sea, and the Great Barrier Reef. They prefer clear and well-oxygenated waters with abundant algae growth, as they primarily feed on various types of algae and marine vegetation. The yellow tang's bright yellow coloration serves as a form of camouflage among the colorful corals and reef structures.

The diet of yellow tangs consists mainly of algae, including filamentous, foliose, and coralline algae species. They are herbivorous grazers, using their small, scraping teeth to feed on algae attached to rocks, coral surfaces, and other substrates within their habitat. Yellow tangs play a vital role in reef ecosystems by helping to control algae growth and maintain the health of coral reefs.

Behaviorally, yellow tangs are social and active swimmers that form schools or shoals in their natural habitat. They are diurnal, meaning they are most active during the day, feeding and interacting with other fish in the reef community. Yellow tangs often exhibit shoaling behavior, where they swim closely together in coordinated movements, providing safety in numbers and enhancing their ability to find food and avoid predators.

One interesting aspect of yellow tangs is their ability to change coloration based on environmental factors and social interactions. They may become paler or darker in response to changes in light intensity, water quality, or social status within the shoal. This color variability allows yellow tangs to adapt to their surroundings and communicate with other fish in their group.

Another fascinating fact about yellow tangs is their breeding behavior and reproductive biology. During the breeding season, which typically occurs in the spring and summer months, male yellow tangs engage in courtship displays to attract females. Once a pair forms, they release eggs and sperm into the water, where fertilization occurs externally. The fertilized eggs develop into larvae, which drift in ocean currents before settling on suitable substrate to grow into juvenile fish.

Yellow tangs are highly valued in the aquarium trade due to their vibrant coloration and peaceful temperament. However, overcollection for the aquarium trade has led to concerns about the sustainability of wild populations. Efforts are being made to promote responsible and sustainable aquaculture practices to reduce the impact on wild yellow tang populations.

In conclusion, the yellow tang is a colorful and ecologically important species in coral reef ecosystems. Its physical appearance, habitat preferences, diet, behavior, and role in reef health make it a subject of interest and conservation efforts among marine enthusiasts and researchers. Protecting and preserving coral reef habitats is crucial for ensuring the continued survival and well-being of yellow tang populations and the diverse marine life they support.

Angelfish

The marine angelfish is a fascinating and visually striking species commonly found in tropical and subtropical waters around the world. These fish belong to the family Pomacanthidae and are known for their vibrant colors, unique markings, and graceful swimming behavior. Marine angelfish come in a variety of species, each with its own distinctive appearance and characteristics.

One of the defining features of marine angelfish is their physical appearance, which includes a compressed and oval-shaped body, elongated fins, and a small mouth equipped with sharp teeth. They are typically brightly colored, with patterns of stripes, spots, or bars that serve as camouflage among coral reefs and rocky substrates. The vibrant colors of marine angelfish, such as yellow, blue, orange, and black, make them popular among aquarium enthusiasts and divers.

Marine angelfish are commonly found in coral reef habitats, including shallow lagoons, reef slopes, and rocky outcrops. They prefer areas with abundant coral growth, as they rely on corals for shelter, food, and breeding sites. Marine angelfish are territorial and often establish and defend specific territories within their reef environment, using aggressive displays and vocalizations to ward off intruders and protect their nesting sites.

The diet of marine angelfish primarily consists of algae, small invertebrates, and planktonic organisms found in their reef habitat. They are omnivorous feeders, consuming a variety of prey items and algae to meet their nutritional needs. Marine angelfish play a vital role in reef ecosystems by helping to control algae growth, clean coral surfaces, and maintain the health of coral reefs.

Behaviorally, marine angelfish are active swimmers with a graceful and darting motion. They use their fins and streamlined bodies to navigate through complex reef structures, darting in and out of crevices and coral formations in search of food and shelter. Marine angelfish are diurnal, meaning they are most active during the day, feeding and interacting with other fish in their reef community.

One interesting aspect of marine angelfish is their complex social structure and breeding behavior. They often form pairs or small groups within their reef habitat, engaging in courtship displays and rituals during the breeding season. Once a pair forms, they release eggs and sperm into the water, where fertilization occurs externally. The fertilized eggs develop into larvae, which drift in ocean currents before settling on suitable substrate to grow into juvenile fish.

Another fascinating fact about marine angelfish is their ability to change coloration and markings based on environmental factors, social interactions, and mood. They may become darker or lighter in response to changes in light intensity, water quality, or social status within their group. This color variability allows marine angelfish to blend in with their surroundings and communicate with other fish in their community.

Marine angelfish are highly valued in the aquarium trade due to their beauty, coloration, and unique behaviors. However, their collection for the aquarium trade has led to concerns about the sustainability of wild populations. Efforts are being made to promote responsible and sustainable aquaculture practices to reduce the impact on wild marine angelfish populations.

In conclusion, marine angelfish are captivating and important species in coral reef ecosystems. Their physical appearance, habitat preferences, diet, behavior, and role in reef health make them a subject of interest and conservation efforts among marine enthusiasts and researchers.

Pufferfish

The pufferfish is a fascinating marine creature known for its unique appearance, defensive capabilities, and interesting behaviors. These fish belong to the family Tetraodontidae and are characterized by their ability to inflate their bodies into a spherical shape as a defense mechanism. Pufferfish come in various species, each with its own distinctive features and behaviors.

One of the defining features of pufferfish is their physical appearance, which includes a rounded body, small fins, and a distinctive face with eyes set high on the head. They are typically covered in spines or prickles, which are used for protection against predators. Pufferfish can vary in coloration, ranging from shades of yellow, brown, and gray to more vibrant colors such as green, blue, and orange.

Pufferfish are commonly found in tropical and subtropical waters around the world, including coral reefs, coastal areas, and estuaries. They prefer shallow waters with sandy or rocky substrates, where they can find food and shelter. Pufferfish are often encountered in areas with abundant marine life and suitable breeding grounds.

The diet of pufferfish primarily consists of small crustaceans, mollusks, and invertebrates found in their habitat. They are opportunistic feeders, using their strong beaks and teeth to crush and consume hard-shelled prey items. Pufferfish also feed on algae and plant matter, making them omnivorous and adaptable to various food sources.

Behaviorally, pufferfish are known for their unique defense mechanism of inflating their bodies when threatened. When threatened by predators or perceived danger, pufferfish can rapidly inflate their bodies by swallowing water or air, turning themselves into a round and spiky ball that is difficult for predators to swallow or attack. This defensive tactic deters many predators and reduces the risk of predation.

One interesting aspect of pufferfish is their ability to produce and release a toxin called tetrodotoxin, which is found in their skin, organs, and tissues. This toxin is highly potent and can be lethal to predators and humans if ingested in sufficient quantities. Pufferfish use this toxin as a defense mechanism against predators, deterring them from attacking or consuming them.

Another fascinating fact about pufferfish is their courtship and mating behavior. During the breeding season, male pufferfish engage in elaborate courtship displays to attract females. These displays may include colorful body markings, inflated displays, and chasing behaviors. Once a pair forms, they release eggs and sperm into the water, where fertilization occurs externally. The fertilized eggs develop into larvae, which eventually settle on the ocean floor and grow into juvenile pufferfish.

Pufferfish are also known for their intelligence and problem-solving abilities. They are capable of learning and remembering tasks, such as navigating mazes or finding hidden food rewards. Pufferfish have been observed using their beaks and mouths to manipulate objects and solve puzzles, demonstrating their cognitive abilities.

In conclusion, the pufferfish is a fascinating and adaptive species with unique physical features, defensive capabilities, and behaviors. Its ability to inflate, produce toxins, engage in courtship displays, and demonstrate intelligence makes it a subject of interest and study among marine biologists and enthusiasts. Understanding and conserving pufferfish populations are crucial for maintaining the balance of marine ecosystems and protecting these intriguing and valuable marine creatures.

Lionfish

The lionfish is a striking and invasive marine species known for its vibrant colors, venomous spines, and predatory behavior. These fish, also known as zebrafish or firefish, belong to the family Scorpaenidae and are native to the Indo-Pacific region. However, due to accidental or intentional releases into non-native waters, lionfish populations have become established in the Caribbean Sea, Gulf of Mexico, and parts of the Atlantic Ocean.

One of the defining features of lionfish is their physical appearance, which includes elongated dorsal and pectoral fins, bold stripes or patterns, and venomous spines along their back and sides. They are typically brightly colored, with combinations of red, brown, white, and black markings that serve as a warning to potential predators. Lionfish have a streamlined body shape, allowing them to maneuver efficiently through coral reefs and rocky habitats.

Lionfish are commonly found in tropical and subtropical waters with coral reef ecosystems, including shallow reefs, wrecks, and rocky outcrops. They prefer areas with abundant prey items, such as small fish, crustaceans, and invertebrates, which they capture using their rapid strike-and-gulp feeding technique. Lionfish are opportunistic predators and can thrive in a variety of habitats, making them successful invaders in non-native waters.

The diet of lionfish primarily consists of small fish, crustaceans, and invertebrates that inhabit coral reef environments. They are voracious predators, using their large mouths and expandable stomachs to consume prey items whole. Lionfish are ambush hunters, relying on stealth and camouflage to approach and capture their prey with quick strikes of their venomous spines.

Behaviorally, lionfish are solitary and territorial predators that establish and defend specific areas within their reef habitat. They use their venomous spines as a defense mechanism against potential threats, including predators and competitors. Lionfish are known for their slow and deliberate swimming style, often hovering near coral formations or rocky crevices while waiting for prey to come within striking distance.

One interesting aspect of lionfish is their venomous spines, which contain a potent toxin that can cause painful stings and health complications in humans. The venomous spines serve as a deterrent to predators and provide lionfish with protection against potential threats. It is important for divers and beachgoers to exercise caution and avoid contact with lionfish spines to prevent stings and injuries.

Another fascinating fact about lionfish is their rapid reproductive rate and ability to colonize new habitats. Female lionfish can produce thousands of eggs in a single spawning event, releasing them into the water where fertilization occurs externally. The fertilized eggs develop into larvae, which drift in ocean currents before settling on suitable substrate to grow into juvenile lionfish. This reproductive strategy allows lionfish populations to rapidly increase and expand their range in non-native waters.

Lionfish are also known for their impact on native ecosystems and biodiversity. As invasive species, lionfish can outcompete native fish species for food and habitat, disrupting the balance of marine ecosystems. Their voracious appetite and rapid reproductive rate contribute to their success as invaders, posing challenges for conservation efforts and marine management strategies.

In conclusion, the lionfish is a visually striking but problematic species in marine ecosystems. Its physical appearance, habitat preferences, diet, behavior, and impact as an invasive species make it a subject of interest and concern among marine biologists and environmentalists.

Parrotfish

The parrotfish is a fascinating and colorful marine species known for its distinctive beak-like mouth, vibrant colors, and important role in coral reef ecosystems. These fish belong to the family Scaridae and are found in tropical and subtropical waters around the world. Parrotfish come in a variety of species, each with its own unique coloration, size, and habitat preferences.

One of the defining features of parrotfish is their physical appearance, which includes a beak-shaped mouth with fused teeth, a robust body, and bright colors ranging from shades of blue, green, yellow, and pink. They are named for their beak-like mouth, which they use to scrape algae and coral polyps from rocks and coral reefs. Parrotfish are also known for their ability to change coloration as they mature, with juveniles often exhibiting different colors than adults.

Parrotfish are commonly found in coral reef habitats, including shallow reefs, lagoons, and rocky outcrops with abundant coral growth. They play a crucial role in reef ecosystems by grazing on algae and dead coral, which helps to prevent algae overgrowth and promote coral health. Parrotfish are diurnal feeders, meaning they are most active during the day and spend their time foraging for food among coral formations.

The diet of parrotfish primarily consists of algae, coral polyps, and other small invertebrates found in their reef environment. They use their powerful jaws and beak-like teeth to scrape algae and small organisms from rocks and coral structures. Parrotfish are herbivorous grazers, consuming large quantities of algae and organic matter to meet their nutritional needs.

Behaviorally, parrotfish are social and territorial fish that form shoals or schools within their reef habitat. They establish and defend specific feeding territories, using aggressive displays and vocalizations to ward off intruders and protect their food resources. Parrotfish are also known for their rhythmic swimming patterns and darting movements, which they use to navigate through complex reef structures.

One interesting aspect of parrotfish is their unique reproductive biology and mating behavior. During the breeding season, which varies depending on the species and location, male parrotfish engage in courtship displays to attract females. These displays may include colorful body markings, fin displays, and chasing behaviors. Once a pair forms, they release eggs and sperm into the water, where fertilization occurs externally. The fertilized eggs develop into larvae, which eventually settle on suitable substrate to grow into juvenile parrotfish.

Another fascinating fact about parrotfish is their ability to produce mucus cocoons as a defense mechanism against predators and parasites. Parrotfish secrete a protective mucus from their skin, which forms a cocoon-like structure around their bodies while they sleep at night. This mucus cocoon helps to mask their scent and deter predators from detecting them while they rest.

Parrotfish are also known for their role in coral reef formation and bioerosion. As they feed on algae and scrape coral structures, parrotfish produce fine sand particles as a byproduct. Over time, these sand particles accumulate and contribute to the formation of coral sand beaches and sediments. Additionally, the bioerosion process conducted by parrotfish helps to break down dead coral and promote the growth of new coral colonies.

In conclusion, the parrotfish is a visually stunning and ecologically important species in coral reef ecosystems. Its physical appearance, habitat preferences, diet, behavior, and unique adaptations make it a subject of interest and study among marine biologists and enthusiasts. Protecting and conserving parrotfish populations is crucial for maintaining the health and diversity of coral reef habitats worldwide.

Crustaceans

Lobster

The lobster is a fascinating marine crustacean known for its distinctive appearance, culinary significance, and unique behaviors. These creatures belong to the family Nephropidae and are commonly found in coastal waters around the world. Lobsters come in various species, each with its own size, coloration, and habitat preferences.

One of the defining features of lobsters is their physical appearance, which includes a robust body with a hard exoskeleton, jointed legs, and large claws called chelae. They are typically colored in shades of green, brown, or blue, with mottled patterns that provide camouflage among rocky seabeds and underwater structures. Lobsters have compound eyes and sensitive antennae that they use for sensing their environment and detecting food.

Lobsters are commonly found in coastal and offshore waters with rocky or sandy bottoms, where they can find suitable hiding places and food sources. They prefer habitats with abundant marine life, such as rocky reefs, kelp forests, and underwater caves. Lobsters are nocturnal feeders, meaning they are most active during the night and spend their days hiding in crevices or burrows.

The diet of lobsters primarily consists of small fish, crustaceans, mollusks, and marine vegetation found in their habitat. They are opportunistic feeders, using their powerful claws and mouthparts to capture and crush prey items. Lobsters are scavengers and will consume a variety of food sources, including carrion and organic matter.

Behaviorally, lobsters are solitary and territorial creatures that establish and defend specific areas within their habitat. They use their claws and aggressive displays to deter intruders and protect their territory. Lobsters are also known for their ability to communicate through chemical signals, such as releasing pheromones to attract mates or signal aggression.

One interesting aspect of lobsters is their molting process, where they shed their old exoskeleton and grow a new one. Lobsters undergo molting several times throughout their lives, with frequency decreasing as they reach maturity. During molting, lobsters are vulnerable to predation and must find a safe hiding place to protect themselves while their new exoskeleton hardens.

Another fascinating fact about lobsters is their longevity and growth rate. Lobsters can live for several decades, with some species reaching ages of 50 years or more. Their growth rate is slow but steady, with larger lobsters often being older and more dominant within their population. Lobsters continue to grow throughout their lives, molting periodically to accommodate their increasing size.

Lobsters are also known for their social hierarchy and mating behavior. Male lobsters engage in aggressive displays and fights to establish dominance and access to breeding females. Once a dominant male secures a mate, they engage in a courtship ritual that involves the male using its claws to grasp the female's carapace. The female then releases eggs, which are fertilized externally by the male.

In conclusion, the lobster is a fascinating and important species in marine ecosystems. Its physical appearance, habitat preferences, diet, behavior, and unique adaptations make it a subject of interest and study among marine biologists and fisheries experts. Lobsters play a crucial role in marine food webs and are valued for their economic and cultural significance in seafood markets and culinary traditions. Protecting and managing lobster populations is essential for maintaining healthy marine ecosystems and sustainable fisheries.

Hermit Crab

The marine hermit crab is a fascinating crustacean species known for its unique adaptations, including its shell-dwelling behavior and scavenging diet. These crabs belong to the family Paguridae and are commonly found in coastal and marine habitats around the world. Marine hermit crabs come in various species, each with its own size, coloration, and habitat preferences.

One of the defining features of marine hermit crabs is their physical appearance, which includes a soft, curved abdomen that fits into a borrowed shell for protection. They have jointed legs and claws, with one claw being larger and more robust than the other. Marine hermit crabs are typically colored in shades of brown, red, or orange, blending in with their surroundings to avoid predators.

Marine hermit crabs are commonly found in intertidal zones, rocky shores, and coral reefs, where they can find suitable shells and hiding places. They prefer areas with abundant marine life and substrate materials, such as sandy or gravelly bottoms. Marine hermit crabs are nocturnal and spend their days hidden in shells or crevices, emerging at night to forage for food.

The diet of marine hermit crabs primarily consists of algae, detritus, small invertebrates, and carrion found in their habitat. They are opportunistic scavengers, using their claws and mouthparts to scavenge and consume organic matter. Marine hermit crabs play a crucial role in nutrient cycling and ecosystem maintenance by recycling nutrients and cleaning up decaying matter.

Behaviorally, marine hermit crabs are solitary creatures that establish and defend territories within their habitat. They are known for their shell-dwelling behavior, where they use empty gastropod shells as portable homes. Marine hermit crabs can change shells as they grow, seeking out larger or more suitable shells to accommodate their increasing size.

One interesting aspect of marine hermit crabs is their shell-swapping behavior, known as "housing exchange." When a hermit crab outgrows its current shell, it seeks out a larger shell and engages in a competitive interaction with other crabs to claim the new shell. This housing exchange can involve physical battles and challenges between crabs to determine hierarchy and access to shells.

Another fascinating fact about marine hermit crabs is their ability to adapt to a wide range of environmental conditions. They are found in various marine habitats, including tidal pools, estuaries, and deep-sea environments. Marine hermit crabs can withstand fluctuations in temperature, salinity, and oxygen levels, making them highly adaptable to changing conditions.

Marine hermit crabs are also known for their symbiotic relationships with other marine organisms. They often host small anemones or algae on their shells, providing protection and camouflage while gaining benefits from the symbiotic partners. These relationships can enhance the survival and well-being of marine hermit crabs in their natural habitat.

In conclusion, the marine hermit crab is a fascinating and adaptable species with unique behaviors and adaptations. Its physical appearance, habitat preferences, diet, behavior, and symbiotic relationships make it a subject of interest and study among marine biologists and naturalists. Protecting and preserving marine habitats is crucial for ensuring the continued survival and well-being of marine hermit crab populations and the diverse marine life they support.

Decorator Crab

The decorator crab is a fascinating crustacean known for its unique ability to camouflage itself by attaching various objects and materials to its body. These crabs belong to the family Majidae and are found in coastal and marine habitats worldwide. Decorator crabs come in different species, each with its own size, coloration, and decorative preferences.

One of the defining features of decorator crabs is their physical appearance, which includes a rounded body covered with small hairs and spines. They have jointed legs and claws that they use for walking and grasping objects. Decorator crabs are typically colored in shades of brown, green, or red, providing them with a base color to blend in with their environment.

Decorator crabs are commonly found in shallow coastal waters, rocky reefs, and seagrass beds, where they can find suitable objects for decoration and hiding places. They prefer areas with abundant marine life and substrate materials, such as algae, sponges, and coral fragments. Decorator crabs are nocturnal and spend their days hidden among rocks or vegetation, emerging at night to search for food and decorations.

The diet of decorator crabs primarily consists of algae, small invertebrates, and detritus found in their habitat. They are opportunistic feeders, using their claws and mouthparts to scavenge and consume organic matter. Decorator crabs are also known for their herbivorous tendencies, grazing on algae and plant material to supplement their diet.

Behaviorally, decorator crabs are known for their unique camouflage and defensive strategies. They use their specialized hairs and spines to attach objects and materials to their body, such as algae, sponges, shells, and even small animals. This camouflage helps them blend in with their surroundings and avoid detection by predators.

One interesting aspect of decorator crabs is their ability to change decorations based on their environment and predators. They can selectively attach or detach objects from their body to match the colors and textures of their surroundings. This adaptive camouflage allows decorator crabs to evade predators and enhance their chances of survival.

Another fascinating fact about decorator crabs is their role in marine ecosystems and nutrient cycling. As they graze on algae and organic matter, decorator crabs contribute to the breakdown of nutrients and recycling of nutrients in their habitat. They play a crucial role in maintaining the health and balance of coastal and reef ecosystems.

Decorator crabs are also known for their social behavior and interactions with other marine organisms. They often form aggregations or groups within their habitat, where they engage in cooperative feeding and grooming behaviors. Decorator crabs may also engage in territorial displays and interactions with other crabs to establish dominance and access to resources.

In conclusion, the decorator crab is a fascinating and adaptive species with unique camouflage and behavioral traits. Its physical appearance, habitat preferences, diet, behavior, and ecological roles make it a subject of interest and study among marine biologists and naturalists. Protecting and preserving marine habitats is crucial for ensuring the continued survival and well-being of decorator crab populations and the diverse marine life they support.

King Crab

The king crab is a formidable marine crustacean renowned for its large size, powerful claws, and culinary value. These crabs belong to the family Lithodidae and are primarily found in cold, deep-sea environments, particularly in the waters of the North Pacific Ocean. King crabs are known for their robust appearance, which includes a broad carapace, long legs, and formidable claws that earned them their regal name.

In terms of physical appearance, king crabs are notable for their impressive size, with some species reaching lengths of over a meter from claw to claw. They have a distinctively shaped carapace, which is the hard shell covering their body, and long, spindly legs that allow them to navigate rocky and uneven seabeds. The most striking feature of king crabs is their massive claws, which are used for defense, crushing prey, and manipulation of objects in their environment.

King crabs are typically found in cold, deep-sea habitats, including continental shelves, submarine canyons, and underwater ridges. They prefer areas with rocky or sandy bottoms where they can find shelter and suitable prey. King crabs are well-adapted to the frigid temperatures and high pressure conditions of deep-sea environments, making them resilient and successful in their habitat.

The diet of king crabs is varied and includes a range of marine organisms such as fish, mollusks, crustaceans, and detritus. They are opportunistic scavengers and predators, using their powerful claws and mouthparts to crush and consume prey items. King crabs are also known to feed on carrion and organic matter that drifts down from the surface.

Behaviorally, king crabs are solitary creatures that are primarily active at night. During the day, they often hide in crevices or burrows to avoid predators and conserve energy. King crabs are known for their slow and deliberate movements, using their sensory antennae to detect vibrations and chemical cues in the water to locate food and potential mates.

One interesting aspect of king crabs is their molting process, where they shed their old exoskeleton to grow a new one. Molting is a vulnerable time for king crabs as they are soft-bodied and susceptible to predation. They often seek sheltered areas during molting to protect themselves until their new exoskeleton hardens.

Another fascinating fact about king crabs is their economic importance and value in the seafood industry. They are prized for their succulent meat, particularly their large claws, which are a delicacy in many cuisines around the world. King crabs are commercially harvested for their meat, with regulated fisheries to ensure sustainable management of populations.

King crabs also play a role in marine ecosystems as scavengers and predators. They help to maintain the balance of marine food webs by consuming dead and decaying matter, recycling nutrients, and controlling populations of smaller organisms. Additionally, king crabs are preyed upon by larger marine predators such as fish, seals, and sea otters, contributing to the complex interactions within marine ecosystems.

In conclusion, the king crab is a remarkable and valuable species in marine environments. Its physical appearance, habitat preferences, diet, behavior, and ecological roles make it a subject of interest and importance in marine biology and fisheries management. Conservation efforts are essential to ensure the sustainable harvest and preservation of king crab populations for future generations.

Shrimp

Shrimp are small, fascinating crustaceans that inhabit a wide range of aquatic environments worldwide, from freshwater rivers and lakes to coastal seas and deep oceans. They belong to the order Decapoda and are characterized by their slender, elongated bodies, segmented exoskeletons, and distinctive curved tails. Shrimp come in various species, each with its unique colors, sizes, and adaptations to their specific habitats.

These creatures are well-suited to a diverse array of environments, including coral reefs, estuaries, and muddy seabeds. Their habitats often provide them with ample hiding spots among rocks, vegetation, or burrows in the substrate, allowing them to evade predators and hunt for food efficiently. Shrimp are also found in commercial aquaculture operations, where they are bred and raised for human consumption.

In terms of diet, shrimp are omnivorous scavengers, feeding on a wide range of organic matter and small organisms. They consume algae, detritus, plankton, small fish, and invertebrates like mollusks and worms. Shrimp use their specialized mouthparts, including mandibles and maxillipeds, to grasp, manipulate, and crush food items before ingesting them.

Behaviorally, shrimp exhibit fascinating social behaviors and mating rituals. They often form schools or shoals, particularly during their juvenile stages, to protect themselves from predators and improve foraging efficiency. Shrimp are also known for their intricate courtship displays, which involve visual cues such as color changes, tail flicking, and chemical signals released into the water to attract potential mates.

One interesting fact about shrimp is their ability to regenerate lost limbs. If a shrimp loses a claw or leg due to predation or injury, it can regenerate the missing appendage through a process called molting. During molting, shrimp shed their old exoskeleton and grow a new one, including any lost limbs, allowing them to continue their normal activities relatively unimpeded.

Another fascinating aspect of shrimp is their role in marine ecosystems as keystone species. They serve as important prey for a variety of predators, including fish, seabirds, and marine mammals, contributing to the transfer of energy and nutrients within food webs. Additionally, shrimp play a vital role in nutrient cycling and water filtration, as they consume organic matter and algae, helping to maintain water quality and ecosystem balance.

Shrimp also have unique adaptations for survival, including their ability to camouflage and avoid detection by predators. Some shrimp species can change color to match their surroundings, using pigment cells called chromatophores in their skin to blend in with the substrate or vegetation. This camouflage helps them avoid detection and increases their chances of survival in their habitat.

In addition to their ecological importance, shrimp have significant economic value as a seafood commodity. They are harvested and cultivated for human consumption, both in commercial fisheries and aquaculture operations. Shrimp are consumed in various culinary dishes worldwide and are prized for their delicate flavor and nutritional content, including high protein and essential nutrients like vitamins and minerals.

In conclusion, shrimp are fascinating creatures with diverse adaptations, behaviors, and ecological roles. Their physical appearance, habitat preferences, diet, and behavior make them a subject of interest in marine biology, aquaculture, and fisheries management. Understanding and conserving shrimp populations is essential for maintaining healthy marine ecosystems and sustainable seafood resources for future generations.

Krill

Krill are small, shrimp-like crustaceans that play a crucial role in marine ecosystems as a primary food source for many marine animals. They belong to the order Euphausiacea and are characterized by their transparent bodies, segmented exoskeletons, and distinctive swimming appendages. Krill are found in large swarms or schools in cold, nutrient-rich waters around the world, particularly in the Southern Ocean and polar regions.

Physically, krill have elongated bodies with multiple segments, including a head, thorax, and abdomen. They have a pair of antennae and several pairs of thoracic limbs, which they use for swimming and feeding. Krill are typically transparent or translucent, allowing them to blend in with their surroundings and avoid detection by predators.

Krill thrive in cold, nutrient-rich waters, such as those found in polar oceans and upwelling zones. They are well-adapted to survive in these environments, where they form dense aggregations or swarms near the surface. Krill feed on phytoplankton, microscopic algae, and organic matter suspended in the water column, using their filtering appendages to strain food particles from the water.

The diet of krill primarily consists of phytoplankton, which they filter from the water using specialized feeding appendages called thoracic limbs or pleopods. Krill are filter feeders, using their appendages to create a current that draws in water and planktonic organisms. They then use fine hairs on their limbs to capture and consume the tiny organisms.

Behaviorally, krill are known for their diel vertical migration, where they move between surface waters at night to feed on phytoplankton and deeper waters during the day to avoid predators. This migration pattern helps krill optimize their feeding opportunities while minimizing their risk of predation by visual predators like fish and birds.

One interesting fact about krill is their role as a keystone species in marine ecosystems. They are a vital food source for a wide range of marine animals, including fish, seabirds, whales, and seals. The abundance of krill in an ecosystem can have cascading effects on predator populations, making them essential for maintaining healthy marine food webs.

Krill also have unique adaptations for survival, including their ability to produce bioluminescent light. Some krill species can emit light from specialized organs called photophores, which they use for communication, camouflage, and predator avoidance in the deep ocean where light is limited.

Another fascinating aspect of krill is their reproductive biology and life cycle. Female krill can produce thousands of eggs, which they release into the water where they hatch into larvae. Krill larvae go through several developmental stages before maturing into adults, during which they undergo molting to shed their old exoskeletons and grow larger.

Krill are also commercially harvested for their high protein and nutrient content, particularly in the aquaculture and fisheries industries. They are processed into krill oil, a valuable source of omega-3 fatty acids and antioxidants used in dietary supplements and pharmaceuticals.

In conclusion, krill are essential organisms in marine ecosystems, with unique physical adaptations, feeding behaviors, and ecological roles. Their abundance and importance as a primary food source make them a subject of interest in marine biology, fisheries management, and conservation efforts. Protecting and managing krill populations is crucial for maintaining healthy oceans and sustainable fisheries for future generations.

Whales

Orca

The orca, also known as the killer whale, is a magnificent apex predator found in oceans worldwide, from the Arctic to Antarctic regions. They are easily recognizable by their distinct black and white coloration, robust bodies, and tall dorsal fins. Orcas belong to the dolphin family, Delphinidae, and are the largest members of this family, with males reaching lengths of up to 30 feet and weighing several tons.

These majestic creatures inhabit a wide range of marine environments, including coastal waters, open oceans, and even some rivers and estuaries. They are highly adaptable and can be found in both cold and warm waters, from polar regions to tropical seas. Orcas are known for their impressive swimming abilities and can reach speeds of up to 34 miles per hour, making them formidable hunters in their environment.

Orcas are carnivorous predators with a diverse diet that includes fish, squid, seals, sea lions, and even other marine mammals like dolphins and whales. They are apex predators, meaning they are at the top of the marine food chain and have no natural predators themselves. Orcas are skilled hunters, using a variety of hunting techniques depending on their prey, including cooperative hunting strategies where they work together in pods to capture larger prey.

Behaviorally, orcas are highly social animals that live in complex social structures known as pods. Pods consist of several individuals, typically led by a dominant female known as the matriarch. Orcas within a pod communicate using a variety of vocalizations, clicks, whistles, and calls, which are used for social bonding, navigation, and hunting coordination.

One interesting fact about orcas is their distinctive coloration and markings, which vary among different populations and ecotypes. While most orcas have black backs, white bellies, and a white patch above their eyes called the eye patch, some populations exhibit unique variations in coloration and markings. For example, some orcas in the Antarctic have a larger eyepatch, while those in the Pacific Northwest may have gray or yellowish coloration on their bodies.

Another fascinating aspect of orcas is their intelligence and problem-solving abilities. They are known for their complex behaviors, including tool use, cooperative hunting, and cultural traditions passed down through generations. Orcas have large, highly developed brains that allow them to learn, adapt, and communicate effectively within their social groups.

Orcas also have a long lifespan, with females living up to 90 years or more and males up to 50-60 years. They reach sexual maturity between 8-15 years of age, and females typically give birth to a single calf every 3-10 years. Calves are born after a gestation period of about 15-18 months and are cared for by their mothers and other pod members.

In addition to their ecological importance, orcas have cultural significance for many indigenous communities around the world. They are revered as symbols of strength, wisdom, and family bonds, and their presence in traditional stories and ceremonies reflects their importance in these cultures.

In conclusion, orcas are magnificent marine mammals with a wide range of physical adaptations, behaviors, and ecological roles. Their striking appearance, diverse diet, social complexity, intelligence, and cultural significance make them a subject of fascination and admiration for people around the world. Protecting and conserving orca populations is crucial for maintaining healthy marine ecosystems and preserving these iconic creatures for future generations.

Blue Whale

The blue whale is an awe-inspiring marine mammal, renowned as the largest animal to have ever lived on Earth. These majestic creatures belong to the baleen whale family, Balaenopteridae, and are characterized by their massive size, streamlined bodies, and distinct bluish-gray coloration. Blue whales can grow up to lengths of 100 feet or more and weigh as much as 200 tons, making them true giants of the ocean.

Blue whales are primarily found in deep ocean waters, including the open seas and polar regions. They have a wide global distribution, with populations inhabiting all major oceans, from the Arctic to the Antarctic. Blue whales prefer areas with rich food sources, such as krill and small fish, which are abundant in cold, nutrient-rich waters.

In terms of diet, blue whales are filter feeders that primarily consume krill, a small shrimp-like crustacean found in dense swarms in the ocean. They use their baleen plates, which are made of keratin, to filter seawater and trap krill while expelling excess water. Blue whales can consume up to 4 tons of krill per day during feeding seasons, making them one of the most efficient feeders in the animal kingdom.

Behaviorally, blue whales are known for their solitary nature and long migratory journeys. They are often seen traveling alone or in small groups, although they may temporarily gather in larger aggregations during feeding or mating seasons. Blue whales are also known for their impressive vocalizations, including deep, low-frequency calls that can travel long distances underwater.

One interesting fact about blue whales is their remarkable size and anatomy. They have a heart that can weigh as much as a car and is the largest known in any animal species. Despite their massive size, blue whales are incredibly agile swimmers, capable of reaching speeds of up to 20 miles per hour when necessary.

Another fascinating aspect of blue whales is their annual migration patterns. They undertake long-distance migrations between feeding and breeding grounds, traveling thousands of miles each year. These migrations are essential for accessing seasonal food sources and for reproductive purposes.

Blue whales also have unique social behaviors, including communication through vocalizations and physical interactions such as breaching, spyhopping, and lobtailing. Breaching, where the whale leaps out of the water and crashes back down, is believed to serve various purposes, including communication, removing parasites, or stunning prey.

In addition to their ecological importance as apex predators, blue whales have historical and cultural significance for many communities around the world. They have been revered in indigenous cultures as symbols of strength, wisdom, and spirituality, and their presence in oceans has inspired awe and admiration throughout human history.

In conclusion, the blue whale is a magnificent marine mammal with a range of physical adaptations, behaviors, and ecological roles. Its massive size, deep-ocean habitat, krill-based diet, migratory patterns, and social behaviors make it a subject of fascination and conservation concern. Protecting and preserving blue whale populations and their habitats is crucial for maintaining healthy marine ecosystems and ensuring the survival of these iconic giants of the sea.

Beluga Whale

The beluga whale is a distinctive and enchanting marine mammal known for its striking appearance and sociable nature. Belugas, also called white whales, are easily recognized by their milky-white skin and unique forehead bulge called a melon. They belong to the cetacean family Monodontidae and are found in Arctic and subarctic waters, including the Bering Sea, Hudson Bay, and the Arctic Ocean.

Physically, beluga whales have a rounded body shape with a small dorsal fin and a flexible neck that allows them to move their heads in various directions. Their white coloration helps them blend in with icy Arctic environments, although some individuals may have spots or patches of gray, particularly as they age. Belugas also have a distinctive melon on their heads, which they use for echolocation and communication.

Belugas inhabit a range of marine environments, including coastal waters, estuaries, and deep ocean regions near ice floes. They are well-adapted to cold, icy waters and are often found in areas with seasonal sea ice cover. Belugas migrate seasonally, moving between feeding and breeding grounds depending on the availability of food and optimal conditions for mating and calving.

The diet of beluga whales is diverse and includes a variety of fish, crustaceans, squid, and other small marine organisms. They are opportunistic feeders, using their flexible necks and sharp teeth to capture and consume prey. Belugas may also use echolocation to locate food in dark or murky waters, emitting high-frequency sounds and interpreting the returning echoes to detect prey.

Behaviorally, beluga whales are highly social animals that live in pods or groups consisting of several individuals, including adults, juveniles, and calves. Pods are often led by older females known as matriarchs, who guide group movements and social interactions. Belugas communicate using a range of vocalizations, clicks, whistles, and squeals, which play a crucial role in navigation, finding mates, and maintaining social bonds within the pod.

One interesting fact about beluga whales is their ability to mimic sounds, including human speech. They have a diverse range of vocalizations and can imitate a variety of sounds, from whistles and clicks to human-like vocalizations. This mimicry has made them popular attractions in marine parks and aquariums, where they delight audiences with their vocal abilities.

Another fascinating aspect of beluga whales is their adaptability to changing environments. They can adjust their behavior and movements in response to environmental changes, such as fluctuations in sea ice cover, water temperature, and prey availability. Belugas are resilient animals that have survived in harsh Arctic conditions for thousands of years.

Belugas also have a unique annual molting process, where they shed their old skin and grow new layers of blubber and epidermis. This molting helps them maintain optimal insulation and buoyancy in cold waters, as well as remove accumulated parasites and dead skin cells.

In addition to their ecological importance as predators and indicators of Arctic ecosystem health, beluga whales have cultural significance for indigenous communities in the Arctic region. They are revered as symbols of wisdom, resilience, and harmony with nature, and their presence in traditional stories, art, and ceremonies reflects their importance in these cultures.

In conclusion, the beluga whale is a captivating and iconic marine mammal with a range of physical adaptations, behaviors, and ecological roles.

Narwhal

The narwhal is an intriguing Arctic whale known for its long, spiral tusk that can grow up to 10 feet in length. These whales, also called the "unicorns of the sea," belong to the Monodontidae family, along with belugas. Narwhals have a sleek, mottled gray or brownish skin coloration that helps them blend into their icy Arctic environment. They are medium-sized whales, reaching lengths of about 13 to 18 feet.

Narwhals are primarily found in Arctic and subarctic waters, including the Canadian Arctic, Greenland, and the Russian Arctic. They prefer deep, offshore waters with seasonal sea ice cover, where they can find their preferred prey and avoid predators. Narwhals are well-adapted to cold environments, with thick blubber layers for insulation and a specialized circulatory system that helps regulate body temperature in icy waters.

The diet of narwhals consists mainly of fish, such as Arctic cod, halibut, and shrimp, along with squid and other small marine organisms. They use their sharp teeth to capture and eat prey, relying on echolocation to navigate and locate food in dark or turbid waters. Narwhals are skilled hunters that can dive to depths of over 3,000 feet in search of food.

Behaviorally, narwhals are known for their social structure and vocalizations. They live in groups called pods, which can range in size from a few individuals to several dozen members. Pods are often segregated by gender and age, with females and calves forming separate groups from males. Narwhals communicate using a variety of vocalizations, clicks, whistles, and buzzes, which play a role in social bonding, navigation, and foraging coordination within the pod.

One interesting fact about narwhals is their unique tusk, which is actually an elongated tooth that protrudes from the left side of their upper jaw. The tusk is spiral-shaped and can grow to lengths of up to 10 feet in males, although some females and juveniles may have smaller or absent tusks. The function of the tusk is not fully understood, but it is believed to play a role in sensory perception, social interactions, and possibly as a tool for hunting or breaking ice.

Another fascinating aspect of narwhals is their ability to perform deep and prolonged dives. They have specialized adaptations for diving, including a collapsible rib cage that allows them to compress their lungs and conserve oxygen during dives. Narwhals can stay submerged for extended periods, up to 25 minutes or more, as they search for prey in the depths of the Arctic Ocean.

Narwhals are also known for their annual migrations between summer feeding grounds and winter breeding areas. During the summer months, they forage in productive Arctic waters, building up energy reserves for the winter. In winter, they migrate to ice-covered areas for breeding and calving, where they form large aggregations in areas of open water or thin ice.

In addition to their ecological importance as predators in Arctic ecosystems, narwhals have cultural significance for indigenous communities in the Arctic region. They are revered as symbols of strength, wisdom, and resilience, and their presence in traditional stories, art, and ceremonies reflects their importance in these cultures.

In conclusion, the narwhal is a fascinating and enigmatic Arctic whale with unique physical adaptations, behaviors, and ecological roles. Its spiral tusk, Arctic habitat, diet, social structure, and diving abilities make it a subject of fascination and conservation interest. Protecting and preserving narwhal populations and their habitats is crucial for maintaining healthy Arctic ecosystems and ensuring the survival of these iconic and mysterious whales for future generations.

Humpback Whale

The humpback whale is a majestic marine mammal known for its massive size, distinct body shape, and acrobatic behaviors. These whales belong to the family Balaenopteridae and are characterized by their long, narrow bodies, distinctively curved dorsal fins, and large flippers. Humpback whales are among the largest animals on Earth, with adults reaching lengths of up to 50 feet and weighing around 30-40 tons.

Humpback whales inhabit a wide range of marine environments, from polar regions to tropical seas, including the Arctic, Antarctic, and many coastal and offshore waters. They are migratory animals that travel thousands of miles each year between feeding and breeding grounds. During the summer months, humpback whales can be found in high-latitude feeding areas where they consume vast amounts of food to build up energy reserves for their long migrations.

The diet of humpback whales primarily consists of small schooling fish, such as herring, mackerel, and anchovies, as well as krill and other planktonic organisms. They are filter feeders, using baleen plates in their mouths to strain food from the water. Humpback whales are skilled hunters that use various feeding techniques, including lunge feeding, bubble netting, and cooperative feeding behaviors.

Behaviorally, humpback whales are known for their acrobatic displays and complex vocalizations. They are highly social animals that live in flexible social structures, including loose aggregations and temporary pods during migration and breeding seasons. Humpback whales communicate using a variety of sounds, including songs, clicks, and whistles, which are used for social interactions, navigation, and mating displays.

One interesting fact about humpback whales is their elaborate songs, which are among the most complex and structured vocalizations of any animal. Male humpbacks are known for their long, haunting songs, which can last for hours and are believed to play a role in mating rituals and social bonding within populations. Each song is unique to a specific population and can change over time.

Another fascinating aspect of humpback whales is their impressive migration patterns. They undertake long-distance migrations between feeding and breeding grounds, traveling thousands of miles each year. These migrations are essential for accessing seasonal food sources and for reproductive purposes, as humpback whales gather in breeding areas to mate and give birth.

Humpback whales are also known for their breaching behavior, where they leap out of the water and crash back down with tremendous force. Breaching is believed to serve various purposes, including communication, social bonding, removing parasites, and stunning prey. Humpback whales are also known to slap their flippers and tails on the water's surface, creating loud splashes and sounds.

In addition to their ecological importance as predators and nutrient recyclers in marine ecosystems, humpback whales have cultural significance for many communities around the world. They are revered as symbols of strength, resilience, and harmony with nature, and their presence in traditional stories, art, and ceremonies reflects their importance in these cultures.

In conclusion, the humpback whale is a magnificent and iconic marine mammal with a range of physical adaptations, behaviors, and ecological roles. Its massive size, migratory patterns, diet, social behaviors, and vocalizations make it a subject of fascination and conservation interest. Protecting and preserving humpback whale populations and their habitats is crucial for maintaining healthy ocean ecosystems and ensuring the survival of these remarkable animals for future generations.

Sperm Whale

The sperm whale is a magnificent deep-sea dweller, renowned for its large size, distinct body shape, and impressive diving abilities. These whales belong to the family Physeteridae and are characterized by their massive heads, long, narrow bodies, and large, square-shaped flippers. Sperm whales are among the largest predators on Earth, with males reaching lengths of up to 60 feet and females slightly smaller.

Sperm whales are found in oceans worldwide, from polar regions to tropical seas, although they prefer deep, offshore waters. They inhabit both pelagic and continental shelf areas, diving to great depths in search of their preferred prey. Sperm whales are highly adapted to deep-sea environments, with specialized physiological and anatomical features that allow them to withstand the pressures of deep dives.

The diet of sperm whales consists mainly of squid, particularly deep-sea species like the giant squid and colossal squid. They are skilled hunters that use echolocation to locate prey in dark and deep waters. Sperm whales have a unique hunting strategy where they dive deep to hunt for squid in the abyssal depths, using their powerful jaws and teeth to capture and consume prey.

Behaviorally, sperm whales are known for their complex social structures and communication. They live in social groups called pods, which can consist of several females and their offspring led by a dominant female, known as the matriarch. Adult males are often solitary or form smaller groups known as bachelor pods. Sperm whales communicate using a variety of clicks, codas, and vocalizations, which play a role in social interactions, foraging coordination, and navigation.

One interesting fact about sperm whales is their large, block-shaped head, which contains a massive spermaceti organ filled with a waxy substance called spermaceti. This organ is believed to play a role in buoyancy control, echolocation, and communication. Sperm whales also have the largest brain of any animal, weighing up to 20 pounds, which is thought to contribute to their intelligence and complex behaviors.

Another fascinating aspect of sperm whales is their impressive diving abilities. They are capable of diving to depths of over 3,000 feet and staying submerged for up to 90 minutes or more. Sperm whales have specialized adaptations for deep dives, including collapsible lungs, increased blood volume, and a thick layer of blubber for insulation and energy storage.

Sperm whales are also known for their unique behavior known as "spermaceti running," where they swim at high speeds near the water's surface, creating a distinctive wake behind them. This behavior is believed to serve various purposes, including communication, social interactions, and possibly as a form of play or exercise.

In addition to their ecological importance as top predators in marine ecosystems, sperm whales have historical and cultural significance for many communities around the world. They have been hunted for centuries for their valuable blubber, oil, and other products, leading to population declines and conservation concerns. Sperm whales are now protected under international agreements, although they still face threats from human activities such as pollution, climate change, and entanglement in fishing gear.

In conclusion, the sperm whale is a remarkable marine mammal with a range of physical adaptations, behaviors, and ecological roles. Its large size, deep-sea habitat, squid-based diet, social structure, and diving abilities make it a subject of fascination and conservation interest. Protecting and preserving sperm whale populations and their habitats is crucial for maintaining healthy ocean ecosystems and ensuring the survival of these iconic and enigmatic creatures for future generations.

Minke Whale

The minke whale is a fascinating marine mammal known for its streamlined body, dark gray or black coloration, and pointed snout. These whales belong to the baleen whale family Balaenopteridae and are one of the smallest of the baleen whales, reaching lengths of about 25-30 feet on average. Despite their relatively small size compared to other whale species, minke whales are powerful swimmers and agile hunters.

Minke whales are found in oceans worldwide, from polar regions to temperate seas, although they prefer cool, nutrient-rich waters. They are often seen in coastal areas, offshore waters, and open ocean environments, where they feed on small schooling fish, krill, and other planktonic organisms. Minke whales are known for their migratory behaviors, traveling between feeding and breeding grounds depending on the season and food availability.

In terms of diet, minke whales are opportunistic feeders that consume a variety of prey, including fish species like herring, mackerel, and capelin, as well as krill and small crustaceans. They use baleen plates in their mouths to filter food from the water, swallowing large amounts of water and then using their tongue to push out excess water while retaining prey in their baleen.

Behaviorally, minke whales are relatively solitary and elusive compared to other whale species. They are known for their fast and erratic swimming patterns, making them challenging to observe and study in the wild. Minke whales are also skilled divers, capable of reaching depths of several hundred feet in search of prey. They use echolocation to navigate and locate food in dark or turbid waters.

One interesting fact about minke whales is their vocalizations, which include clicks, whistles, and pulsed sounds. These vocalizations are used for communication, navigation, and possibly foraging coordination. Minke whales are known for their quiet vocalizations compared to other whale species, making them less detectable to predators and human observers.

Another fascinating aspect of minke whales is their breeding and reproductive behaviors. They have a gestation period of about 10-11 months, after which females give birth to a single calf. Calves are born in warm, shallow waters and are nursed by their mothers for several months before becoming independent. Minke whales reach sexual maturity at around 6-8 years of age, and females typically give birth every 2-3 years.

Minke whales are also known for their relatively long lifespan, with individuals living up to 50 years or more in the wild. They face various threats in their environment, including entanglement in fishing gear, habitat degradation, pollution, and climate change. Conservation efforts are underway to protect minke whale populations and their habitats, including measures to reduce bycatch in fisheries and mitigate human impacts on marine ecosystems.

In conclusion, the minke whale is a fascinating and resilient marine mammal with a range of physical adaptations, behaviors, and ecological roles. Its streamlined body, diverse diet, migratory behaviors, vocalizations, and reproductive strategies make it a subject of scientific interest and conservation concern. Protecting and preserving minke whale populations and their habitats is crucial for maintaining healthy marine ecosystems and ensuring the survival of these charismatic creatures for future generations.

Dolphins

Bottlenose Dolphin

The bottlenose dolphin is a charismatic and intelligent marine mammal known for its playful nature, sleek body, and distinctive dorsal fin. These dolphins belong to the family Delphinidae and are found in oceans worldwide, from coastal waters to offshore habitats. They are highly adaptable and can thrive in a variety of marine environments, including temperate and tropical seas, estuaries, and even some freshwater rivers and lakes.

Physically, bottlenose dolphins have a streamlined body shape with a curved dorsal fin and a beak-like snout, which gives them their characteristic "bottlenose" appearance. They are typically dark gray or black on their back and lighter gray or white on their belly, a coloration pattern known as countershading, which helps camouflage them from predators and prey. Bottlenose dolphins have a robust and muscular body, well-suited for fast swimming and agile movements in the water.

Bottlenose dolphins inhabit a wide range of marine environments, including coastal areas, bays, estuaries, and open ocean waters. They are highly adaptable and can thrive in both shallow and deep waters, depending on the availability of food and suitable habitat. Bottlenose dolphins are social animals that often live in pods or groups, ranging in size from a few individuals to several dozen members.

In terms of diet, bottlenose dolphins are opportunistic feeders that consume a variety of prey, including fish, squid, crustaceans, and small marine organisms. They use echolocation to locate and capture prey, emitting high-frequency sounds and interpreting the returning echoes to detect objects and navigate in their environment. Bottlenose dolphins are skilled hunters that work together in coordinated feeding behaviors, such as herding fish into tight groups or using cooperative hunting techniques.

Behaviorally, bottlenose dolphins are known for their playful and social interactions, both within their pods and with other marine animals, including humans. They are highly intelligent and curious creatures that engage in a variety of behaviors, such as leaping, spinning, tail-slapping, and surfing in the waves. These playful behaviors serve multiple purposes, including social bonding, communication, exercise, and possibly foraging practice.

One interesting fact about bottlenose dolphins is their complex communication system, which includes a wide range of vocalizations, clicks, whistles, and body language. Dolphins use these sounds and gestures to convey information, establish social hierarchies, coordinate group activities, and maintain contact with other pod members. Each dolphin has a unique signature whistle, which may serve as a form of individual identification within the pod.

Another fascinating aspect of bottlenose dolphins is their ability to use tools and solve problems. They have been observed using objects like sponges or seashells to protect their snouts while foraging in sandy or rocky areas, a behavior known as "sponging." Dolphins also demonstrate problem-solving skills in captivity, where they can learn and perform complex tasks, such as retrieving objects, following instructions, and participating in cognitive studies.

Bottlenose dolphins are also known for their long lifespan, with individuals living up to 40-50 years or more in the wild. They face various threats in their environment, including habitat loss, pollution, entanglement in fishing gear, and human disturbance. Conservation efforts are underway to protect bottlenose dolphin populations and their habitats, including measures to reduce human impacts and promote sustainable management of marine resources.

Dusky Dolphin

The dusky dolphin is a captivating marine mammal known for its sleek, slender body and distinctive coloration. These dolphins belong to the family Delphinidae and are found in various oceans and seas around the world, particularly in temperate and cool waters. They are highly social animals that often form large pods or groups, ranging in size from a few individuals to several hundred members.

Physically, dusky dolphins have a streamlined body shape with a long, slender beak and a curved dorsal fin. They are typically dark gray or black on their back, fading to lighter shades of gray or white on their belly, with a characteristic dark stripe running from their eye to their flipper. This coloration pattern helps camouflage them from predators and prey, as well as providing thermal regulation by absorbing and reflecting sunlight.

Dusky dolphins inhabit a wide range of marine environments, including coastal waters, bays, estuaries, and offshore habitats. They are often seen in areas with strong currents and upwelling zones, where nutrient-rich waters attract prey and support marine life. Dusky dolphins are highly adaptable and can thrive in both shallow and deep waters, depending on food availability and suitable habitat.

In terms of diet, dusky dolphins are opportunistic feeders that consume a variety of prey, including fish, squid, crustaceans, and small marine organisms. They use echolocation to locate and capture prey, emitting high-frequency sounds and interpreting the returning echoes to detect objects and navigate in their environment. Dusky dolphins are skilled hunters that work together in coordinated feeding behaviors, such as herding fish into tight groups or using cooperative hunting techniques.

Behaviorally, dusky dolphins are known for their acrobatic displays and playful interactions, both within their pods and with other marine animals. They are highly social creatures that engage in a variety of behaviors, such as leaping, spinning, tail-slapping, and surfing in the waves. These playful behaviors serve multiple purposes, including social bonding, communication, exercise, and possibly foraging practice.

One interesting fact about dusky dolphins is their complex communication system, which includes a wide range of vocalizations, clicks, whistles, and body language. Dolphins use these sounds and gestures to convey information, establish social hierarchies, coordinate group activities, and maintain contact with other pod members. Each dolphin has a unique signature whistle, which may serve as a form of individual identification within the pod.

Another fascinating aspect of dusky dolphins is their reproductive behavior and life cycle. They have a gestation period of about 10-11 months, after which females give birth to a single calf. Calves are born in warm, shallow waters and are nursed by their mothers for several months before becoming independent. Dusky dolphins reach sexual maturity at around 5-7 years of age, and females typically give birth every 2-3 years.

Dusky dolphins are also known for their coordinated swimming and hunting behaviors, particularly when feeding on schools of fish or squid. They use strategic movements and vocalizations to communicate and coordinate their actions, allowing them to efficiently capture prey and avoid predators. Dusky dolphins are fast and agile swimmers, capable of reaching speeds of up to 20 miles per hour in short bursts.

Spinner Dolphin

The spinner dolphin is a captivating marine mammal known for its slender body, distinct coloration, and acrobatic behaviors. These dolphins belong to the family Delphinidae and are found in tropical and subtropical waters around the world, particularly in offshore environments and near coral reefs. They are highly social animals that often form large pods or groups, ranging in size from a few individuals to several hundred members.

Physically, spinner dolphins have a streamlined body shape with a long, slender beak and a small, triangular dorsal fin. They are typically dark gray or black on their back, fading to lighter shades of gray or white on their belly, with a characteristic dark stripe running from their eye to their flipper. This coloration pattern helps camouflage them from predators and prey, as well as providing thermal regulation by absorbing and reflecting sunlight.

Spinner dolphins inhabit a wide range of marine environments, including coastal waters, bays, lagoons, and offshore habitats. They are often associated with areas of high productivity, such as upwelling zones and feeding aggregations, where nutrient-rich waters attract prey and support marine life. Spinner dolphins are highly adaptable and can thrive in both shallow and deep waters, depending on food availability and suitable habitat.

In terms of diet, spinner dolphins are opportunistic feeders that consume a variety of prey, including fish, squid, crustaceans, and small marine organisms. They use echolocation to locate and capture prey, emitting high-frequency sounds and interpreting the returning echoes to detect objects and navigate in their environment. Spinner dolphins are skilled hunters that work together in coordinated feeding behaviors, such as herding fish into tight groups or using cooperative hunting techniques.

Behaviorally, spinner dolphins are known for their acrobatic displays and playful interactions, both within their pods and with other marine animals. They are highly social creatures that engage in a variety of behaviors, such as leaping, spinning, tail-slapping, and surfing in the waves. These playful behaviors serve multiple purposes, including social bonding, communication, exercise, and possibly foraging practice.

One interesting fact about spinner dolphins is their unique spinning behavior, where they leap out of the water and spin multiple times before re-entering the water. This behavior is believed to serve various purposes, including communication, social bonding, and possibly as a form of play or display. Spinner dolphins are among the most agile and acrobatic of all dolphin species, capable of spinning several times in rapid succession.

Another fascinating aspect of spinner dolphins is their vocalizations and communication system. They produce a wide range of sounds, including clicks, whistles, and burst pulses, which are used for social interactions, navigation, and foraging. Spinner dolphins are known for their complex vocalizations during group activities, such as feeding, mating, and traveling.

Spinner dolphins also exhibit interesting social behaviors, including cooperative hunting and group cohesion. They often work together in coordinated feeding activities, using strategic movements and vocalizations to communicate and coordinate their actions. Spinner dolphins are highly social animals that form strong bonds within their pods, which can consist of individuals of all ages and sexes.

In addition to their ecological importance as predators and contributors to marine ecosystems, spinner dolphins have cultural significance for many coastal communities. They are revered as symbols of joy, freedom, and harmony with nature, and their presence in tropical waters is celebrated in various cultural traditions, stories, and artwork.

Australian Snubfin Dolphin

The Australian snubfin dolphin is a fascinating and lesser-known species of marine mammal native to northern Australia. These dolphins are distinct in their appearance, behavior, and habitat preferences compared to other dolphin species.

Physically, Australian snubfin dolphins are characterized by their rounded, "snubbed" forehead, which gives them a unique and recognizable appearance. They have a relatively small and stocky body, short beak, and a small, triangular dorsal fin. Their coloration is typically light gray to pale pinkish-gray on their back, fading to white on their belly. These dolphins have a streamlined body shape, allowing them to navigate through coastal waters and estuaries with ease.

The Australian snubfin dolphin is primarily found in shallow coastal waters, estuaries, and river mouths along the northern coast of Australia, particularly in the waters of Queensland and the Northern Territory. They prefer habitats with calm waters, sandy or muddy bottoms, and abundant food sources. Australian snubfin dolphins are often seen in small groups or pods, consisting of a few individuals to around 10 members.

In terms of diet, Australian snubfin dolphins are opportunistic feeders that consume a variety of prey, including fish, crustaceans, and cephalopods. They use echolocation to locate and capture prey, emitting high-frequency clicks and interpreting the returning echoes to detect objects and navigate in their environment. Australian snubfin dolphins are skilled hunters that work together in coordinated feeding behaviors, such as herding fish into tight groups or using cooperative hunting techniques.

Behaviorally, Australian snubfin dolphins are known for their social and playful interactions within their pods. They engage in behaviors such as breaching, tail-slapping, and leaping, which serve multiple purposes including communication, social bonding, and possibly foraging practice. These dolphins are also known for their curiosity towards boats and human activities, often approaching vessels to investigate or play in their wake.

One interesting fact about Australian snubfin dolphins is their unique social structure and vocalizations. They have a complex communication system that includes a variety of clicks, whistles, and pulsed sounds, which are used for social interactions, navigation, and foraging coordination within their pods. Each dolphin has a distinct vocal signature, allowing individuals to recognize and communicate with each other.

Another fascinating aspect of Australian snubfin dolphins is their breeding and reproductive behaviors. They have a gestation period of about 10-12 months, after which females give birth to a single calf. Calves are nursed by their mothers for several months before becoming independent. Australian snubfin dolphins reach sexual maturity at around 5-8 years of age, and females typically give birth every 2-3 years.

Australian snubfin dolphins are also known for their gentle and docile nature towards humans, making them popular subjects for ecotourism and research. However, they face various threats in their environment, including habitat degradation, pollution, entanglement in fishing gear, and human disturbance. Conservation efforts are underway to protect Australian snubfin dolphin populations and their habitats, including measures to reduce human impacts and promote sustainable management of coastal ecosystems.

In conclusion, the Australian snubfin dolphin is a unique and important species of marine mammal with a range of physical adaptations, behaviors, and ecological roles. Its distinctive appearance, habitat preferences, social behaviors, and vocalizations make it a subject of scientific interest and conservation concern.

White Sided Dolphin

The white-sided dolphin, also known as the white-sided Atlantic dolphin, is a striking marine mammal with a distinctive appearance and fascinating behaviors. These dolphins belong to the family Delphinidae and are found primarily in the North Atlantic Ocean, particularly in temperate and subarctic waters.

Physically, white-sided dolphins are known for their vibrant coloration, featuring a dark gray or black back that transitions to a lighter gray or white belly. They have a sleek and streamlined body shape with a curved dorsal fin, long beak, and pointed flippers. The sides of their bodies are adorned with striking patterns of white, yellow, or light gray patches, which give them their name. These dolphins have a robust build, well-suited for fast swimming and agile movements in the water.

White-sided dolphins inhabit a variety of marine environments, including coastal waters, bays, estuaries, and offshore habitats. They are often seen in areas with strong currents and upwelling zones, where nutrient-rich waters attract prey and support marine life. White-sided dolphins are highly adaptable and can thrive in both shallow and deep waters, depending on food availability and suitable habitat.

In terms of diet, white-sided dolphins are opportunistic feeders that consume a variety of prey, including fish, squid, crustaceans, and small marine organisms. They use echolocation to locate and capture prey, emitting high-frequency clicks and interpreting the returning echoes to detect objects and navigate in their environment. White-sided dolphins are skilled hunters that work together in coordinated feeding behaviors, such as herding fish into tight groups or using cooperative hunting techniques.

Behaviorally, white-sided dolphins are known for their acrobatic displays and playful interactions, both within their pods and with other marine animals. They are highly social creatures that engage in a variety of behaviors, such as leaping, spinning, tail-slapping, and surfing in the waves. These playful behaviors serve multiple purposes, including social bonding, communication, exercise, and possibly foraging practice.

One interesting fact about white-sided dolphins is their vocalizations and communication system. They produce a wide range of sounds, including clicks, whistles, and burst pulses, which are used for social interactions, navigation, and foraging. White-sided dolphins are known for their complex vocalizations during group activities, such as feeding, mating, and traveling.

Another fascinating aspect of white-sided dolphins is their cooperative hunting behaviors and group cohesion. They often work together in coordinated feeding activities, using strategic movements and vocalizations to communicate and coordinate their actions. White-sided dolphins are highly social animals that form strong bonds within their pods, which can consist of individuals of all ages and sexes.

White-sided dolphins also exhibit interesting reproductive behaviors and life history. They have a gestation period of about 10-12 months, after which females give birth to a single calf. Calves are born in warm, shallow waters and are nursed by their mothers for several months before becoming independent. White-sided dolphins reach sexual maturity at around 6-8 years of age, and females typically give birth every 3-4 years.

In addition to their ecological importance as predators and contributors to marine ecosystems, white-sided dolphins have cultural significance for many coastal communities. They are revered as symbols of freedom, grace, and harmony with nature, and their presence in the oceans is celebrated in various cultural traditions, stories, and artwork.

Rays

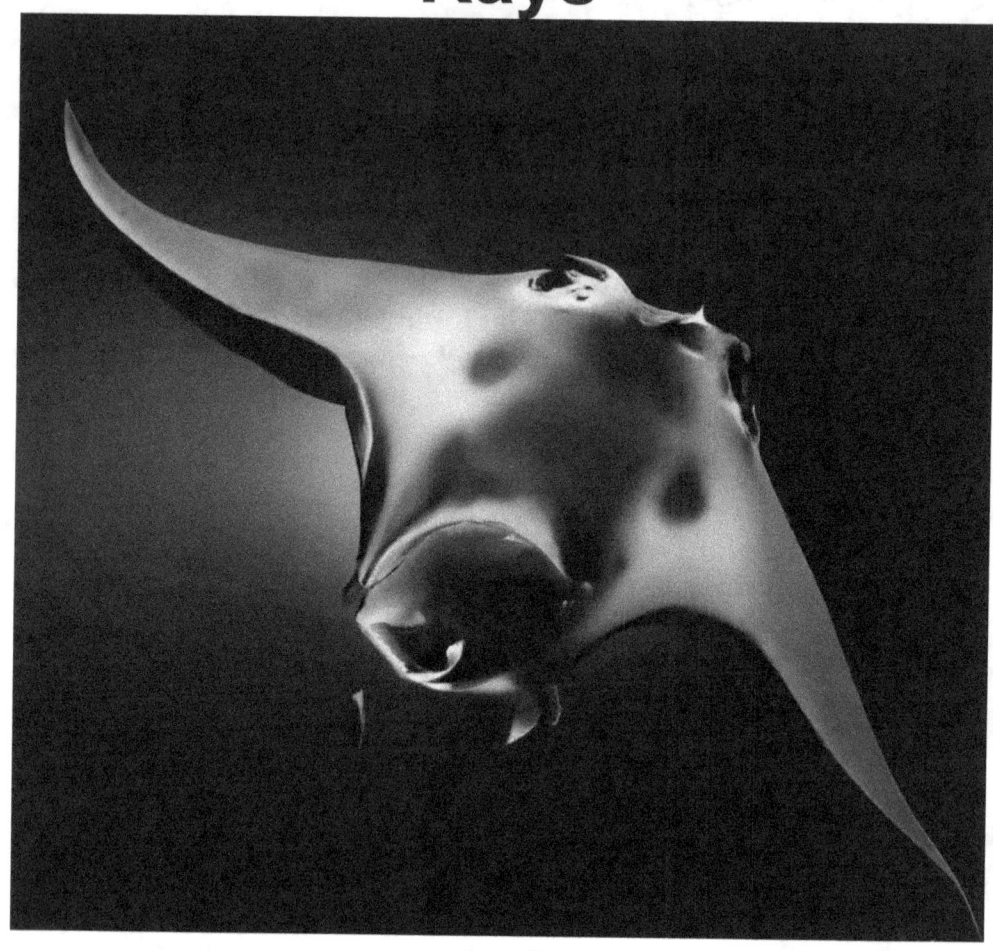

Cownose Ray

The cownose ray is a fascinating marine species known for its unique physical appearance and interesting behaviors. These rays belong to the family Myliobatidae and are found in various coastal and estuarine waters, primarily in the Western Atlantic Ocean, Gulf of Mexico, and Caribbean Sea.

Physically, cownose rays are characterized by their distinctive flattened body shape, which resembles a kite or diamond. They have a broad, rounded head with a pair of large, wing-like pectoral fins that extend from the sides of their body. The upper surface of the cownose ray is typically brown or grayish-brown in color, while the underside is white or pale in hue. These rays have a long, whip-like tail with one or more venomous barbs near the base, which they use for defense against predators.

Cownose rays are well-adapted to a variety of marine environments, including shallow coastal waters, bays, estuaries, and sandy or muddy seabeds. They are commonly seen in areas with abundant seagrass beds and shellfish populations, which provide essential food sources and suitable habitat for these rays. Cownose rays are known to migrate seasonally, often moving in large schools along coastal areas during warmer months.

In terms of diet, cownose rays are primarily carnivorous and feed on a variety of prey, including mollusks, crustaceans, small fish, and benthic invertebrates. They use their specialized dental plates to crush and consume hard-shelled prey items, such as clams, oysters, and crabs. Cownose rays are bottom-dwelling feeders that use their wing-like fins to stir up sediment and uncover buried prey items.

Behaviorally, cownose rays are known for their graceful swimming patterns and social behaviors within their schools. They often move in coordinated groups, gliding through the water with effortless agility using their undulating wing movements. Cownose rays are also capable of leaping out of the water, a behavior known as breaching, which may serve various purposes including communication, predator avoidance, and play.

One interesting fact about cownose rays is their reproductive strategy, which involves internal fertilization and ovoviviparous reproduction. Females give birth to live young, known as pups, after a gestation period of around 11-12 months. The pups are born with a fully formed set of teeth and are immediately capable of swimming and foraging on their own. Cownose rays typically give birth to one or two pups per litter, although litter sizes can vary.

Another fascinating aspect of cownose rays is their migratory behavior and seasonal movements along coastal areas. They are known to undertake long-distance migrations, traveling in large schools or aggregations to specific breeding or feeding grounds. These migrations are often influenced by water temperature, food availability, and reproductive cycles.

Cownose rays play an important ecological role in marine ecosystems as both predators and prey. They help control populations of benthic invertebrates and contribute to nutrient cycling in coastal habitats. Additionally, cownose rays are an important food source for larger predators such as sharks, dolphins, and sea birds.

Despite their ecological significance, cownose rays face various threats in their environment, including overfishing, habitat degradation, pollution, and bycatch in fisheries. Conservation efforts are underway to protect cownose ray populations and promote sustainable management of their habitats. Public awareness and education about the importance of these rays in marine ecosystems are also key components of conservation initiatives.

Giant Reef Manta Ray

The giant reef manta ray is a majestic and captivating marine species known for its impressive size, graceful movements, and unique ecological role in coral reef ecosystems. These rays belong to the family Mobulidae and are found in tropical and subtropical waters around the world, particularly in areas with rich coral reef habitats.

Physically, the giant reef manta ray is one of the largest species of rays, with a wingspan that can exceed 20 feet (6 meters) and a weight of up to 3,000 pounds (1,360 kilograms). They have a flattened body shape, wide pectoral fins that resemble wings, and a distinctive cephalic lobes or "horns" on either side of their mouth. The upper surface of the giant reef manta ray is typically dark brown or black in color, while the underside is lighter in hue.

Giant reef manta rays are primarily found in warm, tropical waters near coral reefs, seamounts, and underwater structures where planktonic food sources are abundant. They are known to inhabit both shallow coastal areas and deeper offshore environments, depending on seasonal variations in water temperature and food availability. Giant reef manta rays are highly migratory and may travel long distances in search of food and suitable breeding grounds.

In terms of diet, giant reef manta rays are filter feeders that consume large quantities of planktonic organisms, including small fish, crustaceans, and microscopic zooplankton. They use their wide mouths and specialized gill rakers to filter plankton from the water as they swim gracefully through the ocean. Giant reef manta rays are known to perform "feeding spirals" where they swim in circular patterns to concentrate and consume planktonic prey.

Behaviorally, giant reef manta rays are known for their gentle and docile nature, often approaching divers and snorkelers in a curious manner. They are agile swimmers that can glide effortlessly through the water using their powerful pectoral fins. Giant reef manta rays are also known for their acrobatic displays, including breaching, somersaulting, and leaping out of the water, which may serve various purposes such as communication, play, or predator avoidance.

One interesting fact about giant reef manta rays is their unique coloration patterns, which are used for individual identification and scientific research. Each ray has a distinct pattern of spots and markings on its ventral surface, similar to a fingerprint, which allows researchers to track and monitor individual rays over time. This information is valuable for studying their movements, behaviors, and population dynamics.

Another fascinating aspect of giant reef manta rays is their reproductive biology and life history. They have a relatively slow reproductive rate, with females giving birth to a single pup every 1-3 years after a gestation period of around 12-13 months. The pups are born live and fully developed, capable of swimming and foraging shortly after birth. Giant reef manta rays exhibit parental care, with females nursing their young and providing protection until they are independent.

Giant reef manta rays play a vital ecological role in coral reef ecosystems as top predators and nutrient cyclers. They help control populations of planktonic organisms, regulate food webs, and contribute to the health and resilience of coral reef communities. Additionally, giant reef manta rays are an important flagship species for marine conservation, drawing attention to the importance of protecting coral reefs and marine biodiversity.

Despite their ecological significance, giant reef manta rays face various threats in their environment, including overfishing, habitat degradation, pollution, and entanglement in fishing gear. Conservation efforts are underway to protect giant reef manta ray populations and their habitats.

Spotted Eagle Ray

The spotted eagle ray is a captivating marine species known for its distinctive appearance, graceful movements, and unique ecological role in coastal and offshore ecosystems. These rays belong to the family Myliobatidae and are found in tropical and subtropical waters worldwide, particularly in areas with coral reefs, seagrass beds, and sandy bottoms.

Physically, the spotted eagle ray is characterized by its flattened body shape, which resembles a diamond or kite, and its long, whip-like tail. They have a wide wingspan that can reach up to 10 feet (3 meters) across and a length of about 5-8 feet (1.5-2.4 meters). The upper surface of the spotted eagle ray is typically dark gray or brown with white spots or rings, while the underside is white or pale in color. They have a pointed snout and distinctive protruding cephalic lobes or "horns" near their eyes.

Spotted eagle rays are primarily found in warm, tropical waters near coastal areas, coral reefs, and estuaries. They prefer habitats with sandy or muddy bottoms, seagrass meadows, and shallow coastal shelves where they can forage for food and seek shelter. Spotted eagle rays are also known to inhabit deeper offshore waters and occasionally venture into brackish or freshwater environments.

In terms of diet, spotted eagle rays are opportunistic feeders that consume a variety of prey, including small fish, crustaceans, mollusks, and benthic invertebrates. They use their specialized dental plates to crush and consume hard-shelled prey items, such as crabs, clams, and oysters. Spotted eagle rays are bottom-dwelling feeders that use their wing-like fins to stir up sediment and uncover buried prey items.

Behaviorally, spotted eagle rays are known for their graceful swimming patterns and acrobatic displays. They are agile swimmers that can glide effortlessly through the water using their powerful pectoral fins. Spotted eagle rays are also capable of leaping out of the water, a behavior known as breaching, which may serve various purposes including communication, predator avoidance, and play.

One interesting fact about spotted eagle rays is their unique reproductive strategy, which involves internal fertilization and ovoviviparous reproduction. Females give birth to live young, known as pups, after a gestation period of around 10-12 months. The pups are born with a fully formed set of teeth and are immediately capable of swimming and foraging on their own. Spotted eagle rays typically give birth to one or two pups per litter, although litter sizes can vary.

Another fascinating aspect of spotted eagle rays is their social behavior and group dynamics. They are often seen in small groups or aggregations, particularly during mating and feeding activities. Spotted eagle rays communicate with each other using body language, gestures, and possibly vocalizations, although their vocal capabilities are not well understood.

Spotted eagle rays play an important ecological role in marine ecosystems as both predators and prey. They help control populations of benthic invertebrates, including crabs and mollusks, and contribute to nutrient cycling in coastal habitats. Additionally, spotted eagle rays are an important food source for larger predators such as sharks, dolphins, and sea birds.

Despite their ecological significance, spotted eagle rays face various threats in their environment, including overfishing, habitat degradation, pollution, and entanglement in fishing gear. Conservation efforts are underway to protect spotted eagle ray populations and promote sustainable management of their habitats. Public awareness and education about the importance of these rays in marine ecosystems are also key components of conservation initiatives.

Round Stingray

The round stingray is a fascinating marine species known for its unique appearance and interesting behaviors. These rays belong to the family Urotrygonidae and are found in coastal waters of the Western Atlantic Ocean, ranging from North Carolina to Brazil, including the Gulf of Mexico and Caribbean Sea.

Physically, the round stingray is named for its circular or rounded body shape, which distinguishes it from other stingray species. They have a flattened disc-like body with a rounded pectoral fin disc that lacks distinct angular corners. The upper surface of the round stingray is typically brown, gray, or olive in color, often with darker mottling or patterns, while the underside is lighter in hue. They have a short, thick tail with one or more venomous spines near the base, which they use for defense against predators.

Round stingrays are primarily found in shallow coastal waters, estuaries, bays, and lagoons with sandy or muddy bottoms. They prefer habitats with soft substrates where they can bury themselves partially to conceal from predators and ambush prey. Round stingrays are also known to inhabit seagrass beds, coral reefs, and rocky areas where they can find food and suitable shelter.

In terms of diet, round stingrays are opportunistic feeders that consume a variety of prey, including small fish, crustaceans, mollusks, and benthic invertebrates. They use their flattened dental plates to crush and consume hard-shelled prey items, such as crabs, clams, and snails. Round stingrays are bottom-dwelling feeders that use their pectoral fins to stir up sediment and uncover buried prey items.

Behaviorally, round stingrays are generally docile and non-aggressive toward humans, although they may sting if threatened or provoked. They are nocturnally active, often foraging for food and moving about their habitats during nighttime hours. Round stingrays are well-adapted for bottom-dwelling life, using their flattened bodies and camouflage to avoid detection by predators and prey alike.

One interesting fact about round stingrays is their reproductive biology and life cycle. They have a unique method of reproduction known as aplacental viviparity, where embryos develop inside the mother's body without a placenta. Female round stingrays give birth to live young, known as pups, after a gestation period of around 4-6 months. The pups are born fully developed and capable of swimming and foraging immediately.

Another fascinating aspect of round stingrays is their symbiotic relationships with other marine organisms. They often host small parasitic copepods or isopods on their bodies, which feed on their skin and tissues. Despite being parasites, these organisms do not typically harm the round stingrays and may even provide some benefits by cleaning their skin and removing dead cells.

Round stingrays also play a role in marine ecosystems as both predators and prey. They help control populations of benthic invertebrates and contribute to nutrient cycling in coastal habitats. Additionally, round stingrays are an important food source for larger predators such as sharks, dolphins, and sea birds.

Despite their ecological significance, round stingrays face various threats in their environment, including habitat loss, pollution, overfishing, and accidental capture in fishing gear. Conservation efforts are underway to protect round stingray populations and promote sustainable management of their habitats. Public awareness and education about these unique marine creatures are also important for fostering conservation action and responsible stewardship of marine resources.

Giant Stingray

The giant stingray, also known scientifically as Himantura chaophraya, is an impressive marine species that inhabits freshwater and brackish environments in Southeast Asia, particularly in the Mekong River basin. Its physical appearance and ecological role make it a fascinating subject for study.

Physically, the giant stingray lives up to its name, as it can grow to immense sizes, with some individuals reaching up to 16 feet (5 meters) in width and weighing over 1,300 pounds (600 kilograms). This ray has a distinctively flattened body with a disc-like shape, allowing it to maneuver gracefully through the water. Its dorsal side is typically dark brown or gray, often with a mottled pattern for camouflage, while the ventral side is lighter in color. The giant stingray is equipped with a long, whip-like tail that ends in a venomous spine, which serves as a defense mechanism against predators.

The environment in which the giant stingray thrives is typically freshwater or brackish, such as large rivers, estuaries, and inland lakes. It prefers areas with sandy or muddy substrates where it can bury itself partially to ambush prey and avoid detection. The Mekong River, with its diverse habitats and abundant food sources, provides an ideal home for these massive rays.

As for its diet, the giant stingray is a carnivorous predator that feeds on a variety of aquatic organisms. Its diet includes fish, crustaceans, mollusks, and benthic invertebrates. The ray uses its electroreceptors, located on its flattened body, to detect the electrical impulses of its prey, allowing it to locate and capture food efficiently.

Behaviorally, the giant stingray is a solitary and elusive creature. It spends much of its time concealed in sandy or muddy substrates, waiting patiently for potential prey to come within striking distance. When hunting, the ray uses its powerful pectoral fins to glide stealthily through the water, relying on its camouflage and sensory adaptations to remain undetected by both prey and predators.

One interesting fact about the giant stingray is its reproductive biology. Like other stingrays, it exhibits aplacental viviparity, where embryos develop inside the mother's body without a placenta. Female giant stingrays give birth to live young, known as pups, after a gestation period of around 9-12 months. These pups are miniature versions of adults and must fend for themselves from birth.

Another fascinating aspect of the giant stingray is its cultural significance in Southeast Asia. In regions where it is native, such as Thailand, Laos, and Cambodia, the giant stingray holds symbolic value and is often featured in local folklore, art, and traditional practices.

Despite its formidable appearance, the giant stingray is generally non-aggressive toward humans and will only use its venomous spine in self-defense if threatened or provoked. However, caution should always be exercised when encountering these powerful creatures in their natural habitat.

Conservation of the giant stingray is of increasing concern due to habitat degradation, overfishing, and dam construction in freshwater ecosystems. Efforts are underway to protect and monitor populations of these rays, as they play a crucial role in maintaining the ecological balance of their habitats.

In conclusion, the giant stingray is a remarkable and enigmatic species that commands attention with its immense size, unique adaptations, and vital role in freshwater ecosystems. Studying and conserving these magnificent creatures is essential for preserving biodiversity and ensuring the health of our planet's freshwater environments.

Turtles

Leatherback Turtle

The leatherback turtle, scientifically known as Dermochelys coriacea, is a remarkable marine reptile renowned for its distinctive physical characteristics, ecological significance, and fascinating behaviors. This species holds a unique place in the world of turtles due to its large size, specialized adaptations, and global distribution across oceans.

Physically, the leatherback turtle is the largest of all sea turtle species, with adults often reaching lengths of 6 to 7 feet (1.8 to 2.1 meters) and weighing up to 2,000 pounds (900 kilograms). Unlike other sea turtles, which have hard bony shells, the leatherback's shell is composed of a leathery, rubbery skin with prominent ridges or keels. Its shell coloration ranges from dark gray to black, and it lacks the typical scutes found in other turtle species. This unique shell structure allows the leatherback to dive to great depths and withstand immense pressure in oceanic environments.

The leatherback turtle is well adapted to a wide range of marine environments, from tropical and subtropical waters to colder temperate regions. It can be found in oceans worldwide, including the Atlantic, Pacific, and Indian Oceans, as well as the Mediterranean Sea. These turtles are highly migratory and can travel thousands of miles between nesting and foraging grounds.

In terms of diet, the leatherback turtle is primarily a carnivorous feeder with a specialized diet of jellyfish and other soft-bodied invertebrates. Its esophagus is lined with backward-pointing spines called papillae, which help prevent jellyfish tentacles from harming the turtle's digestive tract. Leatherbacks are known to consume large quantities of jellyfish, using their powerful jaws and throat muscles to swallow their gelatinous prey whole.

Behaviorally, leatherback turtles are known for their impressive diving abilities and long-distance migrations. They can dive to depths of over 4,000 feet (1,200 meters) and stay submerged for extended periods, thanks to their unique physiological adaptations, such as high blood oxygen levels, efficient oxygen storage, and reduced metabolic rates during dives. These turtles undertake epic migrations between nesting beaches and foraging grounds, traveling thousands of miles across oceanic waters.

One interesting fact about leatherback turtles is their remarkable ability to regulate body temperature in cold water environments. Unlike other sea turtles, which rely on external sources of warmth, leatherbacks have specialized adaptations, such as a thick layer of insulating fat and a unique blood circulation system, which allow them to maintain body heat even in frigid waters.

Another fascinating aspect of leatherback turtles is their nesting behavior and reproductive biology. Females return to their natal beaches every 2 to 3 years to lay their eggs in sandy nests. Leatherback nests are distinctive in shape, with deep, flask-like cavities dug by the female using her powerful flippers. After laying her eggs, the female covers the nest with sand and returns to the sea, leaving the eggs to incubate for around 60 to 70 days before hatching.

Leatherback turtles play a crucial ecological role in marine ecosystems as top predators and nutrient cyclers. By consuming large quantities of jellyfish, they help control populations of these gelatinous organisms, which can otherwise bloom and disrupt marine food webs. Additionally, leatherbacks are preyed upon by sharks, killer whales, and other large predators, contributing to the balance of oceanic ecosystems.

Green Turtle

The green turtle, scientifically known as Chelonia mydas, is a captivating marine reptile famous for its striking appearance, important ecological role, and fascinating behaviors. These turtles are named for the green coloration of their fat and cartilage, not their shells, which can range from olive to brownish-black in hue.

Physically, green turtles have a distinctive oval-shaped carapace, or shell, which can grow up to 5 feet (1.5 meters) in length and weigh several hundred pounds. Their carapace is typically smooth and heart-shaped, with a serrated edge near the rear. The coloration of their carapace can vary depending on factors such as age, diet, and habitat. Juvenile green turtles often have darker shells with prominent patterns, while adults tend to have lighter-colored shells with fewer markings.

Green turtles inhabit a wide range of marine environments, including tropical and subtropical waters around the world. They can be found in coastal areas, coral reefs, seagrass beds, and oceanic habitats, where they feed on a variety of plant and animal matter.

In terms of diet, green turtles are primarily herbivorous, feeding mainly on seagrasses, algae, and marine plants. They use their powerful jaws and beak-like mouths to crop and ingest seagrass blades, which they digest using specialized stomach chambers that house symbiotic bacteria to break down plant cellulose. Green turtles are essential herbivores in marine ecosystems, helping to maintain the health of seagrass beds and coral reefs through their grazing activities.

Behaviorally, green turtles are known for their impressive swimming abilities, long migrations, and unique nesting behaviors. They are strong swimmers capable of traveling long distances between foraging and nesting grounds, often crossing entire ocean basins during migration. Green turtles undertake epic journeys to return to their natal beaches to lay eggs, following ancient migratory routes passed down through generations.

One interesting fact about green turtles is their remarkable navigational skills and homing instincts. Despite traveling thousands of miles across open ocean during migration, female green turtles are able to return with pinpoint accuracy to the exact beach where they were born to nest. This navigational feat remains a mystery to scientists and is thought to involve a combination of celestial cues, magnetic fields, and olfactory cues.

Another fascinating aspect of green turtles is their reproductive biology and nesting behavior. Female green turtles typically nest at night, digging deep pits in sandy beaches with their flippers and laying clutches of 100 to 200 eggs. After covering the nest with sand, the female returns to the sea, leaving the eggs to incubate for around 60 days before hatching. Hatchlings emerge from the nest and make a perilous journey to the water, facing numerous predators along the way.

Green turtles play a crucial ecological role in marine ecosystems as both herbivores and prey. They help control seagrass growth, promote nutrient cycling, and provide food for predators such as sharks, crocodiles, and seabirds. Additionally, green turtles contribute to beach ecosystems through their nesting activities, which help to redistribute nutrients and support dune stabilization.

Conservation of green turtles is a global priority due to their threatened status and ongoing threats such as habitat loss, pollution, climate change, and bycatch in fishing gear. Efforts are underway to protect nesting beaches, reduce human impacts on marine habitats, implement sustainable fishing practices, and monitor green turtle populations to ensure their long-term survival.

Flatback Turtle

The flatback turtle, scientifically known as Natator depressus, is a unique and lesser-known species of marine turtle that inhabits the waters of northern Australia and Papua New Guinea. This turtle species stands out for its distinctive physical features, specialized habitat preferences, feeding habits, behaviors, and intriguing facts.

Physically, the flatback turtle is characterized by its relatively flat and wide carapace, or shell, which is olive to dark brown in color. Unlike some other sea turtles, the flatback's shell lacks distinct ridges or patterns, giving it a smoother appearance. Adult flatback turtles typically measure around 3 to 3.5 feet (90 to 105 centimeters) in carapace length and weigh between 200 to 250 pounds (90 to 115 kilograms). Their flippers are paddle-shaped, adapted for efficient swimming in coastal waters.

The flatback turtle's habitat is primarily restricted to the coastal waters of northern Australia and southern Papua New Guinea. They are commonly found in shallow, nearshore areas with sandy or muddy bottoms, seagrass beds, and coral reefs. These turtles prefer warmer tropical waters and are rarely encountered in colder temperate regions.

In terms of diet, flatback turtles are omnivorous, feeding on a variety of marine organisms depending on their life stage and availability of prey. Their diet includes seagrasses, algae, jellyfish, crustaceans, mollusks, and small fish. Flatbacks are opportunistic feeders, using their strong jaws and beak-like mouths to crush and consume a wide range of food items.

Behaviorally, flatback turtles exhibit nesting behaviors similar to other sea turtle species, with females returning to coastal beaches to lay their eggs. They typically nest at night, using their flippers to dig deep pits in sandy beaches and lay clutches of around 50 to 60 eggs. After covering the nest with sand, the female returns to the sea, leaving the eggs to incubate for approximately 50 to 60 days before hatching.

One interesting fact about flatback turtles is their limited range and specific habitat preferences. Unlike some sea turtle species that have global distributions, flatbacks are endemic to the waters of northern Australia and Papua New Guinea, making them unique to this region.

Another fascinating aspect of flatback turtles is their nesting behavior and reproductive biology. Unlike other sea turtles that may nest multiple times during a nesting season, flatback turtles typically nest only once per season. This nesting strategy may be influenced by the limited availability of suitable nesting beaches in their range.

Flatback turtles are known for their relatively low nesting numbers compared to other sea turtle species. This makes them vulnerable to population declines and highlights the importance of conservation efforts to protect nesting beaches, reduce human impacts, and promote sustainable fishing practices.

Despite their restricted range and lower nesting numbers, flatback turtles play a crucial ecological role in marine ecosystems as herbivores and prey for predators such as sharks and crocodiles. They help maintain the health of seagrass beds and coral reefs through their grazing activities and contribute to nutrient cycling in coastal habitats.

Conservation of flatback turtles is a priority in Australia and Papua New Guinea due to their vulnerable status and ongoing threats such as habitat degradation, pollution, climate change, and accidental capture in fishing gear. Efforts are underway to protect nesting beaches, monitor nesting populations, reduce human impacts, and raise awareness about the importance of conserving these unique marine turtles.

Hawksbill Turtle

The hawksbill turtle, scientifically known as Eretmochelys imbricata, is a fascinating and critically endangered species of marine turtle found in tropical and subtropical waters around the world. This species is renowned for its distinctive physical characteristics, specialized habitat preferences, feeding habits, behaviors, and conservation challenges.

Physically, the hawksbill turtle is distinguished by its unique shell structure, which features overlapping scutes, or plates, that give it a serrated or saw-like appearance. The shell coloration can vary from dark brown to amber, often with streaks of yellow, red, or orange, creating a striking mosaic pattern. Adult hawksbill turtles typically measure around 2 to 3 feet (60 to 90 centimeters) in carapace length and weigh between 100 to 150 pounds (45 to 70 kilograms). Their flippers are paddle-shaped and equipped with claws, making them adept swimmers and climbers.

The hawksbill turtle's habitat is primarily coastal and includes coral reefs, rocky shorelines, mangrove estuaries, and shallow lagoons. They are commonly found in tropical and subtropical waters of the Atlantic, Pacific, and Indian Oceans, where they rely on coral reef ecosystems for feeding, nesting, and shelter.

In terms of diet, hawksbill turtles are omnivorous, with a specialized diet that includes sponges, soft corals, algae, jellyfish, mollusks, and crustaceans. They use their sharp, curved beaks to extract prey from crevices in coral reefs and rock formations, displaying a unique feeding strategy that helps maintain the health of coral reef ecosystems.

Behaviorally, hawksbill turtles are known for their solitary nature, strong swimming abilities, and migratory behavior. They undertake long-distance migrations between foraging and nesting grounds, traveling hundreds or even thousands of miles across oceanic waters. Female hawksbills return to coastal beaches to lay their eggs, often nesting multiple times during the nesting season.

One interesting fact about hawksbill turtles is their role in coral reef ecosystems. As voracious eaters of sponges and other invertebrates, hawksbills help control sponge populations, which can otherwise outcompete corals for space on reefs. This feeding behavior contributes to the health and diversity of coral reef communities.

Another fascinating aspect of hawksbill turtles is their cultural and historical significance. These turtles have been revered and utilized by indigenous cultures for thousands of years, with their shells used in traditional crafts, jewelry, and artistic creations. However, this historical exploitation has contributed to the decline of hawksbill populations and underscores the need for conservation efforts.

Hawksbill turtles face numerous threats in the wild, including habitat loss, pollution, climate change, bycatch in fishing gear, and illegal trade in their shells and products. These factors have led to a dramatic decline in hawksbill populations worldwide, making them critically endangered and requiring urgent conservation action.

Conservation of hawksbill turtles is a global priority, with efforts focused on protecting nesting beaches, reducing human impacts on marine habitats, implementing sustainable fishing practices, and combating illegal wildlife trade. Education and awareness-raising initiatives are also critical to promote conservation awareness and engage local communities in turtle conservation efforts.

In conclusion, the hawksbill turtle is a remarkable and iconic species with unique adaptations, ecological importance, and conservation challenges.

Loggerhead Turtle

The loggerhead turtle, scientifically known as Caretta caretta, is a captivating and widely distributed species of marine turtle found in oceans around the world. This species is recognized for its robust physical characteristics, specialized habitat preferences, feeding habits, behaviors, and conservation significance.

Physically, the loggerhead turtle is characterized by its large and sturdy head, which gives it its name. Adult loggerheads typically have a carapace, or shell, that measures around 3 to 4 feet (90 to 120 centimeters) in length and weighs between 200 to 350 pounds (90 to 160 kilograms). Their carapace is heart-shaped and typically reddish-brown in color, while their flippers are paddle-shaped with claw-like projections.

The loggerhead turtle's habitat encompasses a wide range of marine environments, including coastal waters, coral reefs, seagrass beds, and oceanic habitats. They are commonly found in temperate and tropical regions of the Atlantic, Pacific, and Indian Oceans, where they forage for food and nest on sandy beaches.

In terms of diet, loggerhead turtles are omnivorous, with a varied diet that includes crustaceans, mollusks, fish, jellyfish, seaweed, and sponges. They use their strong jaws and beak-like mouths to crush and consume hard-shelled prey, displaying a feeding strategy adapted to their marine environment.

Behaviorally, loggerhead turtles are known for their long migrations, strong swimming abilities, and nesting behaviors. They undertake epic journeys between foraging and nesting grounds, traveling thousands of miles across oceanic waters. Female loggerheads return to coastal beaches to lay their eggs, typically nesting multiple times during the nesting season.

One interesting fact about loggerhead turtles is their navigational abilities and homing instincts. Despite traveling vast distances during migration, female loggerheads are able to return with precision to the exact beach where they were born to nest. This remarkable navigational feat remains a mystery to scientists and underscores the complexity of sea turtle behavior.

Another fascinating aspect of loggerhead turtles is their role in marine ecosystems as top predators and nutrient cyclers. By consuming a variety of prey species, loggerheads help control populations of invertebrates and maintain the balance of marine food webs. Additionally, their nesting activities contribute to beach ecosystems by redistributing nutrients and promoting dune stabilization.

Loggerhead turtles face numerous threats in the wild, including habitat loss, pollution, climate change, bycatch in fishing gear, and predation by introduced species. These factors have led to population declines and conservation concerns for loggerhead turtles worldwide.

Conservation of loggerhead turtles is a global priority, with efforts focused on protecting nesting beaches, reducing human impacts on marine habitats, implementing sustainable fishing practices, and raising awareness about the importance of turtle conservation. Research and monitoring programs are also essential for understanding loggerhead populations and developing effective conservation strategies.

In conclusion, the loggerhead turtle is a remarkable and resilient species with unique adaptations, ecological importance, and conservation challenges. Protecting and conserving loggerhead populations are essential for maintaining healthy marine ecosystems, preserving biodiversity, and ensuring the survival of these iconic marine turtles for future generations.

Game Fish

Marlin

The marlin is a majestic and powerful species of fish known for its distinctive physical appearance, formidable hunting abilities, and widespread distribution in oceanic waters around the world. These fish belong to the family Istiophoridae and are characterized by their elongated bodies, sharp bills, and striking coloration.

Physically, marlins have streamlined bodies that are built for speed and agility in the water. They are typically large fish, with adult individuals reaching lengths of up to 10 feet (3 meters) or more and weighing several hundred pounds. Their most notable feature is their long, spear-like bill, which is used for stunning and capturing prey during high-speed attacks.

Marlins inhabit a wide range of marine environments, including tropical and subtropical waters of the Atlantic, Pacific, and Indian Oceans. They are commonly found in offshore areas, near continental shelves, oceanic islands, and along major currents where prey abundance is high.

In terms of diet, marlins are apex predators that feed on a variety of prey, including smaller fish, squid, crustaceans, and occasionally even other marine mammals. They use their sharp bills and powerful jaws to slash through schools of fish, stunning and capturing prey with lightning-fast strikes.

Behaviorally, marlins are known for their impressive swimming abilities, often reaching speeds of up to 50 miles per hour (80 kilometers per hour) during pursuit of prey. They are highly migratory and undertake long-distance migrations between feeding and spawning grounds, following oceanic currents and temperature gradients.

One interesting fact about marlins is their role in sport fishing, where they are prized targets for anglers seeking thrilling battles and trophy catches. Marlin fishing tournaments are held worldwide, attracting enthusiasts and professionals who compete to catch the largest and fastest specimens.

Another fascinating aspect of marlins is their courtship and spawning behavior. During the breeding season, male marlins engage in spectacular displays of agility and strength to compete for mating opportunities with females. Mating often occurs near the ocean surface, where fertilized eggs are released into the water to develop into larvae.

Marlins are known for their incredible leaping ability, with some individuals capable of launching themselves out of the water in acrobatic displays known as "tail walks" or "breaches." These aerial maneuvers are believed to serve various purposes, including communication, predator evasion, and stunning prey.

Conservation of marlins is a topic of concern due to overfishing, habitat degradation, and climate change impacts on marine ecosystems. Sustainable fishing practices, habitat protection, and international cooperation are essential for ensuring the long-term survival of marlin populations and maintaining healthy oceanic environments.

In conclusion, the marlin is a fascinating and iconic species of fish with remarkable physical adaptations, predatory behavior, and cultural significance. Studying and protecting marlin populations are crucial for preserving marine biodiversity, supporting sustainable fisheries, and ensuring the continued enjoyment of these magnificent creatures for generations to come.

Swordfish

The swordfish is a formidable and highly recognizable species of fish known for its distinct physical features, powerful hunting abilities, and oceanic habitat preferences. These fish belong to the family Xiphiidae and are characterized by their elongated bodies, distinctive sword-like bills, and impressive size.

Physically, swordfish have streamlined bodies that are built for speed and agility in the water. They are among the largest bony fish species, with adult individuals reaching lengths of up to 10 to 12 feet (3 to 3.5 meters) and weighing several hundred pounds. Their most prominent feature is their long, flat bill, which resembles a sword and gives them their name.

Swordfish inhabit a wide range of marine environments, including temperate and tropical waters of the Atlantic, Pacific, and Indian Oceans. They are commonly found in offshore areas, near continental shelves, oceanic islands, and along major ocean currents where prey abundance is high.

In terms of diet, swordfish are apex predators with a diverse diet that includes smaller fish, squid, crustaceans, and occasionally even other marine mammals. They use their sharp bills and powerful jaws to slash through schools of fish, stunning and capturing prey with rapid strikes.

Behaviorally, swordfish are known for their impressive swimming abilities and stamina, often undertaking long-distance migrations across oceanic waters. They are solitary hunters that patrol the open ocean in search of prey, relying on their keen senses and powerful bodies to navigate and capture food.

One interesting fact about swordfish is their unique hunting strategy involving their sword-like bills. Unlike other fish species that use their mouths to capture prey, swordfish use their bills to slash and injure prey before consuming it. This slashing behavior is believed to be highly effective for stunning and immobilizing fast-moving prey.

Another fascinating aspect of swordfish is their ability to regulate their body temperature, allowing them to thrive in a wide range of oceanic environments. Swordfish are capable of elevating their body temperature above that of the surrounding water, which enhances their hunting and swimming performance.

Swordfish are renowned targets for commercial and recreational fishing due to their large size, fighting abilities, and prized meat. They are sought after by anglers and fishing fleets worldwide, with regulations in place to manage and sustainably harvest swordfish populations.

Conservation of swordfish is a topic of concern due to overfishing, habitat degradation, and bycatch in fishing gear. Sustainable fishing practices, habitat protection, and international cooperation are essential for ensuring the long-term survival of swordfish populations and maintaining healthy oceanic ecosystems.

In conclusion, the swordfish is a remarkable and iconic species of fish with unique physical adaptations, predatory behavior, and cultural significance. Studying and conserving swordfish populations are crucial for preserving marine biodiversity, supporting sustainable fisheries, and ensuring the continued existence of these magnificent creatures for future generations.

Sailfish

The sailfish is an impressive and iconic species of fish known for its striking physical appearance, swift swimming abilities, and vibrant marine habitats. These fish belong to the family Istiophoridae and are distinguished by their elongated bodies, distinctively long dorsal fin, and remarkable speed in the water.

Physically, sailfish have sleek and streamlined bodies that are built for agility and speed. They are among the fastest fish in the ocean, capable of swimming at speeds of up to 68 miles per hour (110 kilometers per hour). The most distinctive feature of sailfish is their elongated dorsal fin, known as a sail, which can be raised or folded down depending on their mood or activity level.

Sailfish are typically found in warm tropical and subtropical waters of the Atlantic, Pacific, and Indian Oceans. They prefer offshore habitats near continental shelves, coral reefs, and major oceanic currents where prey abundance is high. These areas provide sailfish with ample opportunities for hunting and feeding.

In terms of diet, sailfish are voracious predators that primarily feed on small fish, squid, and crustaceans. They use their sharp bills and powerful jaws to slash through schools of prey, stunning and capturing them with lightning-fast strikes. Sailfish are known for their strategic hunting tactics, often working together in groups to herd and trap prey.

Behaviorally, sailfish are highly social and often travel in small groups or schools. They are agile swimmers and use their fins and tails to maneuver swiftly through the water. Sailfish are known for their acrobatic displays, including spectacular leaps and jumps out of the water, which are believed to serve various purposes such as communication, predator evasion, and stunning prey.

One interesting fact about sailfish is their ability to change color and pattern, especially during moments of excitement or aggression. Their normally blue-gray coloration can transform into vibrant hues of purple, silver, and gold, creating a dazzling display of colors along their body and fins.

Another fascinating aspect of sailfish is their courtship and spawning behavior. During the breeding season, male sailfish engage in impressive displays of courtship, including chasing females, raising their sails, and performing synchronized swimming patterns. Mating occurs near the ocean surface, where fertilized eggs are released into the water to develop into larvae.

Sailfish are prized targets for sport fishing enthusiasts due to their impressive size, speed, and fighting abilities. They are known for putting up fierce battles when hooked, often leaping out of the water and making powerful runs to escape capture.

Conservation of sailfish is important to ensure their long-term survival and maintain healthy marine ecosystems. Sustainable fishing practices, habitat protection, and regulations on recreational fishing are essential for managing sailfish populations and preventing overexploitation.

In conclusion, the sailfish is a captivating and charismatic species of fish with remarkable physical adaptations, hunting strategies, and social behaviors. Appreciating and protecting sailfish populations are essential for preserving marine biodiversity, supporting sustainable fisheries, and conserving the natural beauty of our oceans.

Blue Fin Tuna

The bluefin tuna is a magnificent and highly prized species of fish known for its impressive size, swift swimming abilities, and migratory patterns. These fish belong to the family Scombridae and are renowned for their rich flavor, making them a highly sought-after target for commercial and recreational fishing.

Physically, bluefin tuna have sleek and streamlined bodies that are built for endurance and speed. They are among the largest bony fish species, with adult individuals reaching lengths of up to 10 feet (3 meters) or more and weighing several hundred pounds. Their bodies are characterized by a metallic blue coloration on their back and upper sides, while their undersides are silvery white.

Bluefin tuna inhabit a wide range of marine environments, including temperate and tropical waters of the Atlantic, Pacific, and Indian Oceans. They are commonly found in offshore areas, near continental shelves, oceanic islands, and along major ocean currents where prey abundance is high. Bluefin tuna are highly migratory and undertake extensive migrations between feeding and spawning grounds.

In terms of diet, bluefin tuna are voracious predators with a diverse diet that includes small fish, squid, crustaceans, and planktonic organisms. They use their powerful jaws and sharp teeth to capture and consume prey, displaying swift and agile hunting techniques.

Behaviorally, bluefin tuna are known for their incredible swimming abilities and stamina. They are capable of swimming at speeds of up to 40 miles per hour (64 kilometers per hour), allowing them to cover vast distances during migrations and pursuit of prey. Bluefin tuna are also highly social fish, often forming schools or aggregations while foraging or migrating.

One interesting fact about bluefin tuna is their impressive diving capabilities. These fish are capable of diving to depths of over 3,000 feet (900 meters) in search of food, utilizing their streamlined bodies and efficient swimming techniques to navigate the depths of the ocean.

Another fascinating aspect of bluefin tuna is their spawning behavior. During the spawning season, female bluefin tuna release millions of eggs into the water, which are fertilized by male tuna. The fertilized eggs hatch into larvae, which drift in ocean currents until they mature into juvenile tuna and begin their own migratory journeys.

Bluefin tuna are highly valued in commercial fisheries due to their delicious meat, high market demand, and economic importance. However, overfishing and unsustainable fishing practices have led to population declines and conservation concerns for bluefin tuna worldwide.

Conservation of bluefin tuna is a critical issue, with efforts focused on implementing sustainable fishing practices, reducing bycatch in fishing gear, protecting spawning and nursery areas, and regulating international trade. These measures are essential for ensuring the long-term survival of bluefin tuna populations and maintaining healthy marine ecosystems.

In conclusion, the bluefin tuna is a remarkable and iconic species of fish with impressive physical attributes, migratory behaviors, and ecological importance. Preserving and managing bluefin tuna populations are crucial for sustainable fisheries, marine conservation, and the preservation of this magnificent species for future generations.

Tarpon

The tarpon is a magnificent and highly sought-after fish species known for its impressive size, acrobatic leaps, and unique physical features. These fish belong to the family Megalopidae and are often referred to as "silver kings" due to their silvery scales and majestic presence in coastal and estuarine waters.

Physically, tarpon have elongated and robust bodies with large, reflective scales that give them a shimmering appearance. They can grow to impressive sizes, with adult individuals reaching lengths of up to 8 feet (2.5 meters) or more and weighing several hundred pounds. Tarpon are known for their distinctive upward-facing mouths and prominent lower jaws, which are adapted for capturing prey near the water's surface.

Tarpon inhabit a variety of marine and brackish environments, including coastal waters, estuaries, mangrove forests, and shallow bays. They are commonly found in warm tropical and subtropical regions of the Atlantic Ocean, Gulf of Mexico, and Caribbean Sea. Tarpon are known for their ability to tolerate a wide range of salinity levels, allowing them to thrive in diverse habitats.

In terms of diet, tarpon are opportunistic feeders that consume a variety of prey, including small fish, crustaceans, shrimp, and occasionally even birds or small mammals. They are adept hunters, using their powerful jaws and sharp teeth to capture and swallow prey whole. Tarpon are often observed rolling near the surface to feed on baitfish or invertebrates.

Behaviorally, tarpon are known for their acrobatic leaps out of the water, especially when hooked by anglers. These impressive aerial displays can reach heights of several feet and are believed to serve various purposes, including dislodging hooks, evading predators, and stunning prey.

Tarpon are also capable of prolonged and fast-paced swimming, making them formidable predators in their aquatic environment.

One interesting fact about tarpon is their ability to gulp air at the water's surface using their specialized swim bladders. This adaptation allows them to survive in oxygen-poor or brackish waters, including freshwater environments such as rivers, lakes, and coastal marshes.

Another fascinating aspect of tarpon is their migratory behavior, with populations undertaking long-distance movements between feeding and spawning grounds. During the spawning season, tarpon gather in large aggregations near coastal areas, where females release eggs into the water to be fertilized by males. The fertilized eggs hatch into larvae, which drift in ocean currents until they mature into juvenile tarpon.

Tarpon are highly prized targets for sport fishing enthusiasts due to their size, strength, and acrobatic abilities. Anglers from around the world seek the challenge of hooking and landing tarpon, often employing specialized gear and techniques for this pursuit.

Conservation of tarpon is important to ensure their long-term survival and maintain healthy coastal ecosystems. Efforts focused on habitat protection, sustainable fishing practices, and research on tarpon populations are essential for their conservation and the preservation of their natural habitats.

In conclusion, the tarpon is a remarkable and iconic fish species with impressive physical attributes, feeding behaviors, and ecological importance. Appreciating and protecting tarpon populations are crucial for sustainable fisheries, marine conservation, and the enjoyment of recreational fishing opportunities for generations to come.

www.ingramcontent.com/pod-product-compliance
Lightning Source LLC
Chambersburg PA
CBHW052210220526
45471CB00004B/1907